Colloidal Gold: A Ne
for Cytochemical Ma

GW01465154

Julian E. Beesley

Wellcome Research Laboratories
Beckenham
Kent BR3 3BS, UK

Oxford University Press · Royal Microscopical Society · 1989

Oxford University Press, Walton Street, Oxford OX2 6DP

Oxford New York Toronto
Delhi Bombay Calcutta Madras Karachi
Petaling Jaya Singapore Hong Kong Tokyo
Nairobi Dar es Salaam Cape Town
Melbourne Auckland

and associated companies in
Berlin Ibadan

Royal Microscopical Society
37/38 St. Clements
Oxford OX4 1AJ

Oxford is a trade mark of Oxford University Press

Published in the United States
by Oxford University Press, New York

British Library Cataloguing in Publication Data

Beesley, Julian E.
  Colloidal gold.
  1. Medicine. Diagnosis. Use of cell markers
  I. Title   II. Series
  616.07'582
  ISBN 0-19-856418-X

Library of Congress Cataloging in Publication Data

Beesley, Julian E.
  Colloidal gold: a new perspective for cytochemical marking/
  Julian E. Beesley.
      p.   cm.—(Microscopy handbooks; 17)
  Bibliography: p.   Includes index.
  1. Immunogold labeling.   2. Colloidal gold.
  3. Immunocytochemistry—Technique.   I. Title.   II. Series.
  QR187.I482B44 1989   616.07'92'028—dc19   88-27248
  ISBN 0-19-856418-X

Typeset by Cotswold Typesetting Ltd, Gloucester
Printed in Great Britain
at the University Printing House, Oxford
by David Stanford
Printer to the University

Contents

Acknowledgements

I would like to thank the many scientists who have given me helpful advice during our collaborative studies. I would also like to thank my wife Sara for her help with the manuscript and to express my gratitude to Mr R. Suchett-Kaye for carrying out numerous literature searches. Heartfelt thanks must also be given to the staff of the Word Processing Unit, Wellcome Research Laboratories, Beckenham, for their hard work in typing the manuscript.

Introduction to colloidal gold probes

1.1 Introduction

Immunocytochemistry is the use of labelled antibodies as specific reagents for the detection of tissue constituents or antigens *in situ* (Polak and Van Noorden 1987). It is widely used in the biological sciences. Colloidal gold probes are the most recent immunocytochemical marking technique to have been developed for the detection of antigens *in situ*. A colloidal gold probe is a gold sphere, usually between 2 and 40 nm in diameter, which is coated with a selected protein. This probe, in conjunction with a wide range of microscopical and non-microscopical preparative techniques, is used for the localization of specific proteins. Labelling produced by the technique is distinct and of high sensitivity.

1.2 History of colloidal gold marker systems

Colloidal gold cytochemistry was initiated when Faulk and Taylor (1971) reported the adsorption of anti-*Salmonella* rabbit gamma-globulins to colloidal gold spheres which had been prepared by reducing tetrachloroauric acid with an ether solution of white phosphorus. This novel antibody marker was used as a direct immunocytochemical probe for identifying surface antigens on Salmonellae. In the same year Faulk *et al.* (1971) used this immunocolloid method for detecting rhinoviruses coupled to sheep erythrocytes. Frens (1973) soon after published an important paper describing the production of monodisperse spherical particles of colloidal gold of any chosen diameter between 12 nm and 150 nm by reducing gold chloride with various quantities of sodium citrate. An important characteristic of the colloidal gold marker system, that of selecting a specific size of particle for a given purpose, could then be realized. There soon followed a report (Romano *et al.* 1974) of an indirect labelling technique in which an anti-globulin reagent (a secondary antibody) was adsorbed on to the colloidal gold and used to detect primary antibody already bound to antigen. This technique was used (Romano *et al.* 1975) to study the distribution and mobility of the A, D, and C antigens on human red cell membranes. At this time the nature of the interaction between proteins and gold spheres was poorly understood and it appeared, quite erroneously, that not all species of immunoglobulins, notably human and rabbit IgG, formed a stable complex with colloidal gold. Development of the protein A–gold complex overcame this limitation (Romano and Romano 1977) since protein A binds to the Fc region of immunoglobins, such as IgG, from several mammalian species (Roth 1982*a*). Geoghegan and

Ackerman (1977) reported in detail a method, the theory, and the application of adsorption of proteins to colloidal gold, thus removing the previous limitations encountered by Romano *et al.* (1975). Geoghegan and Ackerman (1977) found that adsorption of proteins to gold spheres is pH dependent, the pH conditions correlating with the isoelectric point(s) of the major protein fraction(s); adsorption is also influenced by the interfacial tension, the solubility, and the electrical charge on the molecules. These important discoveries have enabled a wide variety of proteins to be complexed with colloidal gold.

During these formative years of the colloidal gold technique, important reports of lectin–gold conjugates were presented (Bauer *et al.* 1974; Horisberger *et al.* 1975; Horisberger and Rosset 1977) which advanced the immunocolloid technique into both scanning electron microscope technology and double labelling experiments. Lectins are useful proteins because they can recognize fine differences in carbohydrate complex structures (Debray *et al.* 1981). They are widely used in the study of cell surface glycoconjugates (Horisberger 1985).

A major advance in colloidal gold technology occurred when Roth *et al.* (1978, 1980) demonstrated the use and reliability of the protein A–gold technique for antigen localization on ultrathin sections of embedded tissue. This exciting advance provided the ability for the distinct localization of cytoplasmic antigens.

Since these early innovations in colloidal gold technology there has been a continuous deluge of descriptions of new techniques and applications for both cytochemistry and immunocytochemistry, leading to the recent exciting developments for high-resolution light and electron microscope studies, for non-microscopical techniques, and for use in pregnancy diagnostic kits (Organon Teknika NV, Veedijk 58, 2300 Turnhout, Belgium). For this, the gold particles are coated with two different monoclonal antibodies. In the presence of specific antigen the dispersed particles agglutinate in a sandwich-type specific reaction, thereby changing the original reddish purple colour of the reagents to colourless-greyish within a few minutes. The colloidal gold technique has been applied to most areas of biological research and it is now probably the most popular cytochemical marking system in use. This revolution in marking cytochemistry has been fired by the low cost of the technique, the simplicity of the technique, and the high resolution and high yield of information that can be obtained.

1.3 Aims of the book

The intention of this book is to describe the established techniques of gold probe production and gold labelling. Each technique is summarized in a table which can be used without recourse to the text. The original references are also given to enable readers, if they so wish, to trace the origins of each technique.

Antibodies for immunolabelling

2.1 Introduction

The use of high-quality antibodies is essential because colloidal gold probes are so distinct that any unwanted background labelling is immediately apparent. Specific antibodies are therefore crucial for the success of the technique. A good antibody should be of high affinity or binding capacity to its antigen and any unwanted antibodies in the preparation should be of lower affinity than the specific antibody (Polak and Van Noorden 1987). Only a brief introduction to monoclonal and polyclonal antibodies will be presented here and the interested reader is referred to Mason *et al.* (1983). Either monoclonal or polyclonal antibodies may be used for immunolabelling. The choice of antibody is usually governed by availability rather than preferential production of one antibody type.

2.2 Monoclonal antibodies

Monoclonal antibodies may be raised against a wide range of antigens. They are highly homogeneous, usually reacting with only one molecule and with a single antigenic determinant on the molecule. Furthermore, owing to the method of production, monoclonal antibodies can be produced in potentially unlimited amounts (Mason *et al.* 1983). Monoclonal antibodies therefore appear to possess considerable potential in immunocytochemistry.

The high specificity of monoclonal antibodies sometimes causes difficulties in sample preparation for electron immunocytochemistry. Fixation for electron microscopy involves strong cross-linking of amino acids which destroys the tertiary structure of the protein. If the single epitope (antigenic determinant) recognized by a monoclonal antibody is cross-linked, then reaction with the antibody is prevented. This is overcome by using weak reversible fixatives (Chapter 5).

2.3 Polyclonal antibodies

The range of antigens detectable using polyclonal antibodies is limited to those antigenic constituents which can be purified to homogeneity and only a minor proportion of the antibody (less than 20 per cent) is specific to the immunizing antigen. The specific antibodies that are present are highly heterogeneous in antigen-binding affinity and are usually directed against a number of different epitopes on the immunizing antigen. These preparations would be very difficult to standardize from one laboratory to another. Finally, the preparation and purification of satisfactory polyclonal reagents involves considerable expenditure of time and effort (Mason *et al.* 1983). Nonetheless polyclonal antibodies have proved valuable and are still widely used for immunocytochemistry. A significant advantage of polyclonal antibodies for electron microscope immunocytochemistry is that they contain several populations of different antibodies directed to various portions of the antigen molecule. Thus, the chances of the antiserum finding a non-cross-linked portion of the antigen in a fixed preparation are relatively high.

2.4 Antibody testing

Before being used for experimentation, the antibody preparation must be tested for specific antibody. The antiserum may be tested by enzyme-linked immunosorbent assay (ELISA) against the pure antigen, by Western blot analysis, by radioimmunoassay or by being used in immunocytochemical labelling on known positive tissue (Polak and Van Noorden 1987). Although not always possible, the last method is preferable, since it is the test which most closely corresponds to the experimental conditions.

Preparation of protein–gold probes

3.1 Introduction

The manufacture of a gold probe involves two distinct operations. Gold spheres of the required size are formed by the reduction of a gold salt. These spheres are then coated with the chosen protein suspension (Fig. 1). The procedures for producing gold spheres and for complexing these with proteins are similar, whatever the size of gold particle or whichever protein is used.

3.2 Manufacture of the gold colloid

Gold spheres are produced by the chemical reduction of tetrachloroauric acid. They are formed by condensation of metallic gold from a supersaturated solution created by the reduction of Au^{3+} present in the tetrachloroauric acid (Slot and Geuze 1985). Three independent processes have been distinguished in the formation of the colloidal gold sol. The *reduction* of Au^{3+} eventually produces a supersaturated molecular Au solution. As the Au concentration increases *nucleation* is initiated. At this stage the gold atoms cluster and form seeds or nuclei. Further deposition of molecular Au upon the nuclei forms the gold particles, the process described as *particle growth*. Theoretically, the size of the particles is inversely proportional to the cube root of the number of nuclei formed if it is assumed that the conversion of Au^{3+} to Au is complete and the concentration of gold is kept constant (Frens 1973).

If the reduction, and therefore the rate of nucleation, is rapid many small gold particles are formed. This occurs if either white phosphorus, or sodium or potassium thiocyanate is used as the reducing agent. Depending on the exact experimental conditions, gold sols of 2–3 nm or 5–12 nm can be prepared (Roth 1983a; Baschong *et al.* 1985).

If the conditions are such that the reduction of the chloroauric acid is slow, relatively few nuclei are formed and the resulting gold spheres are relatively large. This occurs if sodium citrate is the reducing agent. Some degree of latitude is available, for if the amount of sodium citrate added is varied the size of the gold spheres varies between 15 and 150 nm (Frens 1973).

Gold particles produced by these methods, especially those below 15 nm, are very heterogeneous and therefore must be fractionated by glycerol or sucrose gradient centrifugation before use (Slot and Geuze 1981). Naturally, a homogeneous population is essential if double labelling techniques are required.

These were the routine preparation methods of gold spheres until Slot and

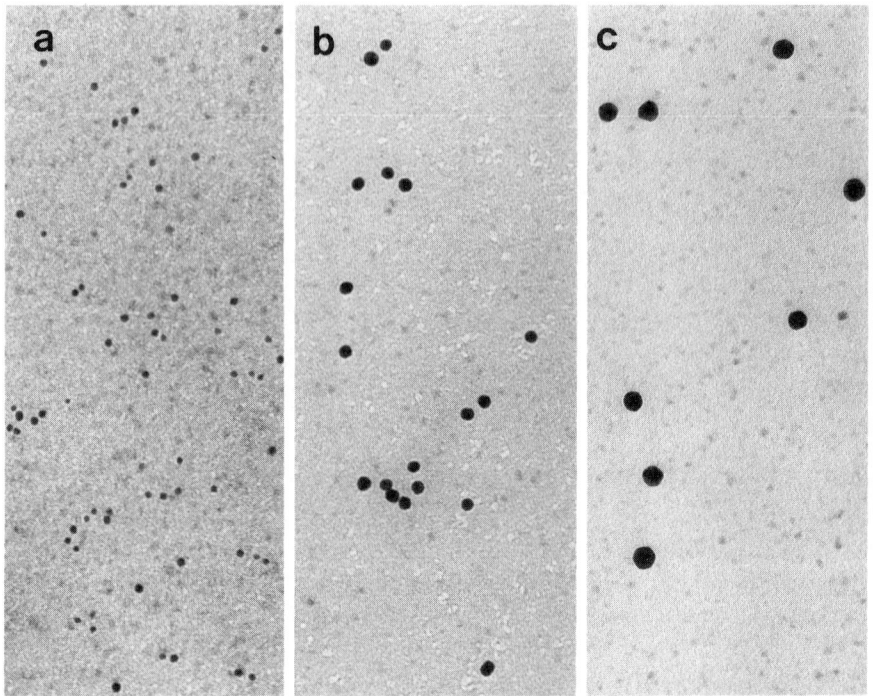

Fig. 1. Colloidal gold probes: 5 nm (a), 10 nm (b), and 15 nm (c) in diameter. These protein–gold complexes can be used over a wide range of magnifications for light and electron microscopy. The probes are distinct and furthermore the obvious size differences allow them to be distinguished easily in multiple labelling experiments. Magnification × 165 000.

Geuze (1985) improved the citrate reduction method for the manufacture of homodisperse gold sols ranging from 3 to 17 nm diameter (Table 3.1). The reduction rate in this technique is controlled by the quantity of tannic acid added, not by the amount of sodium citrate. The technique is feasible only with low-molecular-weight galloylglucose which is a tannic acid prepared from Aleppo nutgalls. The success of this method can be attributed to the combination of reduction rates produced by the sodium citrate and the tannic acid. Sodium citrate induces a slow reduction whilst the reverse occurs with tannic acid. When small particles are required the chemical environment is adjusted so that the reduction is almost entirely effected by the tannic acid. For larger particles insufficient tannic acid is added to reduce all the gold, and thus the sodium citrate completes the reduction. Apparently though, even in this size range, it is still the tannic acid concentration that determines particle size.

The properties of the glass surface are important factors in initiating the reduction process and in determining the reproducibility of the colloidal gold. Small contaminants, such as dust particles cause cloudiness of the colloidal gold, and all glassware used must be thoroughly cleaned and siliconized (Roth 1982a).

Table 3.1. *Preparation of the gold sol (Slot and Geuze 1985)*

1. *The Au^{3+} solution*
   a. 1 ml of 1 per cent HAuCl$_4$
   b. 79 ml distilled water
2. *The reducing mixture*
   a. 4 ml of 1 per cent trisodium citrate 2H$_2$O
   b. 0–2 mls of 1 per cent tannic acid†
   c. 25 mM K$_2$CO$_3$ in the same volume as tannic acid if the tannic acid volume is above 0.5 ml‡
   d. Distilled water to make 20 ml
3. *The reaction*
   i. Warm solutions to 60°C then add the reducing mixture to the Au^{3+} solution quickly and with stirring§
   ii. After the sol has formed heat to boiling, then cool
   iii. Measure at least 100 particles with electron microscope to establish exact size of particles

† 2 ml tannic acid yields 4 nm ± 11.7 per cent particles
 0.5 ml tannic acid yields 6 nm ± 7.3 per cent particles
 0.125 ml tannic acid yields 8.2 nm ± 6.9 per cent particles
 0.03 ml tannic acid yields 11.5 nm ± 6.3 per cent particles
‡ Below 0.5 ml tannic acid has no effect on the pH of the reducing mixture and the K$_2$CO$_3$ may be omitted.
§ At this stage the temperature is critical. The gold sol forms within seconds if a high amount of tannic acid is used. This time increases with lower concentrations of tannic acid, and may be as long as 60 min, if the tannic acid is omitted. Evidence of sol formation is given by the red colour of the mixture.

The use of double-distilled water, filtered before use with a millipore filter system (0.45 μm pore size), for preparation of the different solutions is also advised. Slot and Geuze (1985), however, use the laboratory distilled water and normally cleaned, non-coated glassware. Which approach is used must surely depend on the quality of the laboratory supplies.

3.3 Manufacture of the protein–gold complex
3.3.1 Introduction

Protein adsorption to gold particles is highly complex and still only partially understood. It is assumed that protein adsorption is due to electrostatic interactions between the negatively charged surface of gold particles and positively charged groups of the protein. The protein is attracted on to the action radius of Van der Waals–London attractive forces and firmly bound to the gold particle surface (Roth 1983a). Horisberger and Rosset (1977) observed that neutral polysaccharides, such as yeast mannose and polyethylene glycol, bind to colloidal gold. Roth (1983a) concludes from these observations that there is involvement of other factors, such as electrodynamic attractive forces and adsorption due to the surface roughness of colloidal gold particles in the binding of a protein molecule to a gold sphere. It appears that the molecular orientation of the IgG adsorbed to colloidal gold is dependent on the species, Fab fraction, and pH of absorption (Geoghegan 1985, 1986).

There are several physicochemical factors which affect the adsorption of proteins to gold, but the most important factor, determined by Geoghegan and Ackerman (1977), is the pH at which the adsorption is carried out. The strong adsorption of macromolecules and stable complex formation with gold particles occurs close to or basic to the isoelectric point of the protein, since at these pH values the zwitterion form of the protein is dominant and the interfacial tension is maximal (Roth 1983a). Roth (1983a) reports that this is also the condition under which proteins might more readily adsorb to the hydrophobic surface of gold particles. All reactions involved in complexing proteins with gold spheres are therefore carried out 0.5 pH units above the isoelectric point of the protein.

The binding of a protein to a colloidal gold sphere is a non-covalent electrostatic attraction and it should therefore interfere only minimally with the reactivity of the protein. Handley and Chien (1987), however, believe that the multimeric nature of the gold-labelled ligand needs to be considered in determining the amount of probe to be used in a labelling study, since a probe stabilized with a large number of molecules of a particular ligand is much less effective than the same amount of unlabelled, dispersed molecules in solution. This can be related to the effects of immobilization on the gold surface, to the reduction of active sites of enzymes, and to various forms of steric hindrance. They state that complexing with gold can reduce biological activity of a ligand down to 10 to 20 per cent of the activity in the unlabelled state.

Two processes are involved in the preparation of a gold–protein complex. Firstly, since the proteins involved are often valuable either because so little is available or because they are very expensive, an estimate is made of the minimum amount of protein needed for complete stabilization of the gold sol. In achieving this state, the amount of protein complexed to a gold particle and hence the activity of the probe can be standardized. Finally, these data can be used to manufacture the required amount of gold–protein complex. These techniques may be used to adsorb any protein to the surface of colloidal gold spheres.

3.3.2 Estimating the minimal amount of protein needed to stabilize a gold sol

The stability of the gold sol in water is maintained by repulsion of the electrostatic charges on each gold particle. The addition of strong electrolytes compresses the ion layers around each particle and allows the particles to approach closely to each other. If a critical distance is reached the particles aggregate and flocculation of the sol occurs (Roth 1983a). Proteins adsorbed on the surface of the gold inhibit the electrolyte-induced coagulation of the gold (Roth 1983a). This characteristic is exploited in the form of a simple titration to determine the minimum amount of protein needed to stabilize the gold sol against the flocculating effect of a strong electrolyte.

Flocculation can be judged spectrophotometrically by absorbance at 525–40 nm (Horisberger et al. 1975), flocculation being the point where, if a series of readings are taken, the curve first appears asymptotic with the X-axis (Geoghegan and Ackerman 1977). The maximum absorption varies with the size of the colloid and should be determined for each batch.

The simplest method to judge flocculation, however, is to observe the colour change from red (stabilized gold) to blue (flocculated gold), as described by Roth and Binder (1978). This technique is simple, convenient, and reproducible (Table 3.2).

Table 3.2. *Titration to determine minimum amount of protein to stabilize gold sol (Roth 1982a)*

1. Prepare 10 serial dilutions of the protein in water.
2. Add 0.1 ml of each dilution, with agitation, to a different 0.5 ml aliquot of gold sol adjusted to 0.5 pH units above isoelectric point of the protein.
3. After 1 min at room temperature add 0.1 ml of 10 per cent NaCl, also with agitation.
4. The tube containing the minimum amount of protein needed to stabilize the gold sol can be judged as the one containing the least protein whose sol does not change from red to blue after the addition of the NaCl.

The pH of the colloid is an important factor in these reactions. It is inadvisable to use pH paper for estimating the pH of the sol because this method is not sufficiently accurate. A pH electrode is the preferred equipment but the gold sol will plug the pores unless suitable precautions are taken. Electrode damage is prevented by the addition of a few drops of 1 per cent aqueous polyethylene glycol (MW 20 000) or carbowax-M to a few millilitres of the colloid before measuring the pH (Lucocq and Roth 1985). This stabilized colloid must be discarded after the pH measurement is taken. Alternatively, a gel-filled combination electrode may be used, in which case contact between the gold sol and the electrode should be kept to a minimum.

The pH of the colloid is adjusted by the addition of 0.1 N HCl or acetic acid to lower the pH or by 0.2 M K_2CO_3 to raise the pH (Lucocq and Roth 1985). These authors also suggest that the pH of the colloid may be adjusted by dialysis of the colloid against 2–5 mM buffer solution already adjusted to the desired pH.

3.3.3 Formation of protein–gold complexes

The minimal quantity of protein needed to stabilize a given volume of gold sol is known from the titration described in the previous section. These quantities are scaled up to the required volume. Once the complex has been formed it is then concentrated and purified by suitable ultracentrifugation (Table 3.3–3.7). Proteins most commonly complexed with gold spheres are protein A (Table 3.3) and antibodies, mainly secondary, although primary antibodies are sometimes used (Table 3.4). If biotinylated antibodies are available then the avidin–gold system can be used. A main disadvantage of preparing avidin–gold probes using avidin from egg white is the high isoelectric point, pH 10, of the avidin which may

Table 3.3. *Preparation of the protein A–gold complex (Slot and Geuze 1984)*

1. Take 30 ml of the gold sol.
2. Add sufficient protein A kept as a 2 mg/ml stock solution in distilled water, so that the concentration exceeds the stabilization point by 10 per cent.
3. Add 0.3 ml of 5 per cent polyethylene glycol (Carbowax 20, Union Carbide) or 0.1 per cent bovine serum albumin (Slot and Geuze 1985) after 1 min to ensure that the gold particles are maximally stabilized.
4. Centrifuge to pellet the protein A–gold complex.
   Various conditions have been suggested and these should be evaluated for different preparations and thereafter standardized:

15 nm gold complex 60 000g 1 h at 4°C	(Roth 1982*b*)
12 nm gold complex 5000g 45 min at 4°C	(Slot and Geuze 1984)
5–12 nm gold complex 105 000g 1.5 h at 4°C	(Roth 1982*b*)
5 nm gold complex 125 000g 45 min at 4°C	(Slot and Geuze 1984)
2–3 nm gold 105 000g 1.5 h at 4°C	(Roth 1982*b*)

   The pellet is composed of a large loose part of protein–gold complex and a small tightly packed pellet on the side of the tube. This packed pellet is composed of gold particles that have not been fully stabilized and also aggregated gold particles.
5. The protein–gold complex (the loose pellet) can be resuspended to 1.5 ml in phosphate-buffered (0.01 M, pH 7.4) saline (0.15 M) containing 0.2 mg/ml of polyethylene glycol (Carbowax 20, Union Carbide). This can be stored for up to 1 year at 4°C. It is diluted 10–20 fold before use (Roth 1982*a*).
6. Slot and Geuze (1984) further purify the protein–gold complex by resuspending the complex (the loose pellet) in a small volume of the supernatant and then layering the complex over a 10–30 per cent continuous sucrose or glycerol gradient (volume 10.5 ml, length 8 cm) in phosphate-buffered (0.01 M, pH 7.2) saline (0.15 M). This is centrifuged in a SW41 rotor (Beckman Instruments) for 45 min at 41 000 rpm (5 nm sols) or 30 min at 20 000 rpm (12 nm sols). From about 1 cm below the top down to the bottom, the gradient is stained red by the protein A–gold complex.
7. Collect the upper half of the red zone in successive fractions of 1 ml. These are largely free of clumps and the average sizes should be confirmed by electron microscopy.
8. These probes are diluted as required before use.
9. For long-term storage these authors recommend dialysing the complex to 45 per cent glycerol in phosphate-buffered saline and keeping it at −18°C or by freezing small samples in lower concentrations of glycerol at −70°C.

Table 3.4. *Preparation of antibody–gold complex (Roth 1983a)*

1. Immunoglobulin solutions (1 mg ml^{-1}) are dialysed against 2 mM borax HCl buffer pH 9. They are kept in this low molarity buffer for as short a time as possible.
2. Centrifuge at 100 000g for 1 h at 4°C just before complex formation to sediment any protein aggregates.
3. Adjust pH of colloidal gold to 9 with 0.2 M K_2CO_3 just before use.
4. Determine the amount of protein necessary to stabilize the gold sol, add 10 per cent and for complex formation mix the required amount of protein with the colloidal gold.
5. After 2 min add bovine serum albumin, BSA, (in distilled water, pH 9, adjusted with NaOH and microfiltered) to a final concentration of 1 per cent.
6. Filter with 0.22 μm millipore filter.
7. Centrifuge as for protein A–gold complex, and resuspend pellet in 20 mM Tris-HCl (pH 8.2)–0.15 M NaCl containing 1 per cent BSA (TBS).
8. Repeat centrifugation twice and finally resuspend in TBS containing BSA and NaN_3.

NB Manufacture of monoclonal antibody–gold complexes is identical to this schedule but the pH of the protein and colloidal gold is adjusted to the isoelectric point of the monoclonal antibody.

Table 3.5. *Preparation of the avidin–gold complex (Tolson* et al. *1981)*

1. A range of volumes (10–200 μl) of 0.1 N NaOH are added to separate 1 ml gold sol aliquots.
2. Add 10 μl of 1 mg ml^{-1} avidin in H$_2$O to each gold sol aliquot and whirlimix.
3. 100 μl of 1 per cent saline is added and after 60 min reaction, the suspension is centrifuged at 200 rpm for 5 min.
4. The optimum condition is that requiring the minimum amount of NaOH without causing coarse aggregation (no pellet after centrifugation).
5. Scale up to required volume but this time add 0.5 ml of 1 per cent polyethylene glycol (Carbowax 20, Union Carbide) to the reaction mixture after adding the required amount of avidin (found by the above titration) to discourage aggregation.
6. Ultracentrifugation as described for protein A–gold complex formation.
7. Store in 1 per cent polyethylene glycol (Carbowax 20, Union Carbide) in Tris-glycine buffer pH 11.0 at 5°C.
8. Preparation remains stable for several weeks.

Table 3.6. *Preparation of lectin–gold complex (after Horisberger 1985)*

1. Obtain the optimal pH for preparing the lectin–gold conjugate (for a review see Roth 1983b).
2. Determine the minimal amount of lectin necessary for stabilization of the colloidal gold suspension (Table 3.1).
3. Dissolve the correct amount of lectin in 2 ml water then filter through a millipore filter (0.2–0.4 μm pore size) into an ultraclear plastic bottle.
4. Add required amount of colloid with stirring.
5. After 5 min, add 5 ml of 1 per cent polyethylene glycol (Carbowax 20, Union Carbide) solution and when necessary raise pH to 7–8 with 0.1 M K$_2$CO$_3$.
6. Centrifuge (Table 3.3).
7. Suspend mobile pool of pellet in a suitable buffer such as phosphate-buffered saline at the correct pH containing 0.5 mg/ml polyethylene glycol (Carbowax 20, Union Carbide) or 1 per cent bovine serum albumin to stabilize the colloid.
8. The markers may be centrifuged once more to eliminate traces of unbound lectin or purified further on a continuous sucrose or glycerol gradient (Table 3.3).

Table 3.7. *Preparation of enzyme–gold complex (Bendayan 1985)*

1. Obtain the isoelectric point of the protein.
2. Determine the minimal amount of enzyme necessary for full stabilization of the colloidal gold suspension (Table 3.1).
3. The correct quantity of highly purified enzyme is dissolved in 0.2 ml of bidistilled water in a centrifuge tube and mixed with 10 ml of colloidal gold at the correct pH.
4. Centrifuge at speeds determined by the size of the gold sol (see Table 3.3).
5. As in the preparation of the protein A–gold complex, a dark red loose sediment of the enzyme–gold complex is formed. This is carefully separated from the supernatant which contains free enzyme and from the black firm spot which represents free metallic gold.
6. The enzyme–gold complex is resuspended in 1.5 ml of 0.01 M phosphate-buffered saline (NaH$_2$PO$_4$–Na$_2$HPO$_4$; NaCl, 0.14 M) containing 0.02 per cent polyethylene glycol (Carbowax 20, Union Carbide) brought to the optimal pH for labelling.
7. Enzyme–gold complex is stored at 4°C and used within 10 days of preparation. It is diluted 5–10 fold before use with the phosphate buffer containing 0.02 per cent polyethylene glycol (Carbowax 20, Union Carbide).

cause high non-specific background. Complex formation at this high pH is problematical, since under these conditions colloidal gold tends to aggregate (Roth 1983*a*). Despite this there is a report (Tolson *et al.* 1981) of the production of avidin–gold complexes (Table 3.5). There is now another avidin, streptavidin from bacteria, which is an excellent alternative since the isoelectric point is only pH 6.4 (Roth 1983*a*).

Lectin–gold complexes are becoming increasingly important (Horisberger 1985). The preparation of lectin–gold complexes (Horisberger 1985) is almost identical to the preparation of protein A–gold complexes (Table 3.6).

The enzyme–gold complex (Bendayan 1985) is another potentially useful protein–gold complex (Table 3.7). Specific enzymes, once labelled with an electron-dense marker, permit the direct ultrastructural localization of their corresponding substrates (Bendayan 1985).

The manufacture of protein–gold complexes is now routine. For those who are especially interested in manufacturing protein–gold complexes Janssen Pharmaceutica (Life Science Products, B-2340 Beerse, Belgium) produce an extremely comprehensive booklet entitled *Colloidal gold sols for macromolecule labelling*. The techniques are relatively straightforward and use no complex laboratory equipment. Furthermore, the process does not have to be changed significantly when different proteins are complexed to the gold spheres or when different sized gold spheres are used.

3.3.4 Testing protein–gold complexes

Once colloidal gold probes have been manufactured it is necessary to test them before use in an experimental situation. Bao-le Wang *et al.* (1985) describe a very simple model in which antigens are immobilized on a filter paper strip and labelled with a gold conjugate.

The optimum method for testing probes is, of course, by using the probes in a known positive microscopical system.

3.3.5 Storage of protein–gold complexes

Most authors agree that storage of the protein–gold complexes at 4°C is adequate. Even after one year these appears to be no loss of activity (Roth 1982*b*). Sodium azide (0.5 mg/ml) may be added to the stock suspension if necessary. Alternatively, the small colloids may be filtered through a millipore filter of pore size 0.4 μm (Horisberger 1985) to remove unwanted contaminants.

It has been found that colloid diaylsed to 45 per cent glycerol in phosphate-buffered saline may be stored at -18°C (Slot and Geuze 1984). These authors quote evidence that, in certain cases, 100 μl aliquots of the colloid in low concentrations of glycerol may be stored at -70°C.

If there is doubt about the activity of the colloids after storage, then the acitivty should be tested in a known positive system.

Buffers for immunolabelling

Immunolabelling with colloidal gold probes may be combined with any one of several different specimen preparation techniques to obtain specific antigenic information from the sample. Each technique is similar in principle; the antigen is incubated with specific antibody, then washed and incubated with the gold probe before contrasting and examination (Table 4.1). The correct choice of antibody

Table 4.1. *Generalized immunolabelling schedule (Beesley 1987)*

1. Prepare antigen for immunolabelling and incubate with the following:
2. Phosphate-buffered saline pH 7.2 containing 1 per cent bovine serum albumin (PBS BSA) 5 min.
3. Phosphate-buffered saline (PBS) containing 1 per cent gelatin, 10 min.
4. PBS containing 0.02 M glycine, 3 min.
5. PBS BSA, 2×1 min washes.
6. Antibody diluted with PBS BSA for 1 h.
7. Wash PBS BSA 5×1 min.
8. Gold probe diluted with PBS BSA for 1 h.
9. Rinse in PBS.
10. Fix in 2.5 per cent glutaraldehyde in PBS, 2–10 min.
11. Rinse 5×1 min in distilled water.
12. Treat samples in preparation for viewing by routine technique.

and gold probe dilution is necessary if there is to be optimal immunolabelling with minimal background labelling. The required concentrations are found in each case by testing serial dilutions of the reagents. The buffer used for diluting the reagents is important and must be compatible with the immunological reagents. An ideal buffer system for diluting the reagents and washing the preparations is phosphate-buffered (0.01 M, pH 7.2) saline (0.15 M), containing 1 per cent bovine serum albumin (Slot and Geuze 1984).

It is possible that tissues may posses sites which are 'sticky' for all proteins. These sites need to be blocked before application of the immunological reagents to prevent non-specific labelling. Bovine serum albumin in the buffer reduces non-specific attachment of antibodies by competing with antibody for these non-specific 'sticky' sites. A short pre-incubation of the tissue with 1 per cent gelatin in phosphate buffer also reduces non-immunological sticking of the antibody since the gelatin attaches to the 'sticky' sites. Birrell *et al.* 1987 maintain that cold-water fish gelatin is ideal for reducing this non-specific labelling. Naturally, gold probes do not usually recognize bovine serum albumin or gelatin. Gelatin, if included with the gold probe, attaches to free, active sites on the gold spheres, thereby further preventing the probe from attaching to free receptors on the tissue (Behnke *et al.* 1986).

Free aldehyde groups may remain on the tissue if the tissue has been fixed with aldehyde. These may non-specifically 'fix' antibody to the tissue. These reactive sites are blocked by pretreating the sample with 0.02 M glycine in phosphate-buffered saline before immunolabelling.

These precautions, together with thorough washings, are usually sufficient to prevent unwanted non-immunological immunolabelling. Some samples, however, such as histological sections of abnormal or damaged tissue, whole cells in smears (Polak and Van Noorden 1987), and even some constituents in normal cells prepared for electron microscopy, may bind antibodies non-specifically. Sophisticated blocking reagents and washing procedures are therefore no substitute for thorough control experiments at all times.

Finally, after immunolabelling the sample is washed briefly with phosphate-buffered saline to remove unwanted protein content and then the immunolabelling is 'fixed' to the antigen with buffered glutaraldehyde. This prevents dissociation of the antigen–antibody complex during subsequent washing in distilled water and contrasting (Duerrenberger et al. 1986).

Fixation for immunolabelling

5.1 Fixation for electron microscopy

Several preparation schedules for immunolabelling necessitate fixation of the sample. The type of fixative used is dependent on the type of antibody that will be used for immunolabelling.

Fixation for optimum morphological ultrastructure involves strong cross-linking of amino acids which destroys the tertiary structure of the protein. If the antigenic determinant is cross-linked the reaction with the antibody will be prevented. It is therefore sometimes necessary, to sacrifice morphological preservation in order to retain antigenicity.

Monoclonal antibodies usually react with a single antigenic determinant of a few amino acids on each molecule. Light fixation of the antigen is necessary in order that the important amino acids are not affected by the fixative. The fixative should be a mixture of freshly prepared 4 per cent formaldehyde with 0.05 per cent glutaraldehyde in either phosphate or cacodylate buffer. If this destroys the antigenic determinants within the sample, then 4 per cent formaldehyde only should be used as the fixative. Formaldehyde is a reversible fixative and so the samples should be stored in the fixative until needed. Care must also be taken, for if weak fixation is employed to protect the antigen, it may be possible to reach the state where the antigen is not fixed and it leaches out of the sample during processing.

As previously explained (Section 2.3), polyclonal antibodies contain several populations of different antibodies directed to various portions of the antigen molecule. Fixation in this case is carried out with 1 per cent glutaraldehyde in either phosphate or cacodylate buffer.

Osmium tetroxide (Bendayan and Zollinger 1982) and uranyl acetate (Beesley 1987) have been used as fixatives for immunocytochemistry but these are not recommended for routine use.

5.2 Fixation for light microscopy

Morphological requirements are less stringent for light microscope studies than for electron microscope experiments. This gives a certain latitude with choice of fixative. Sections of unfixed fresh frozen tissue may be used, or these may be post-fixed in acetone or alcohol. Alternatively, the tissue may be prefixed with parabenzoquinone or paraformaldehyde. Freeze-dried tissue may be fixed with formaldehyde or parabenzoquinone vapour (Polak and Van Noorden 1987).

Colloidal gold immunolabelling techniques for electron microscopy

6.1 Introduction

Colloidal gold immunolabelling techniques are exceptionally versatile and may be combined with a number of different specimen preparation techniques to obtain specific antigenic information from the sample. The antigens are prepared and labelled following the generalized labelling schedule (Table 4.1), with slight variations to take into account the characteristics of each sample preparation method. Reagent concentrations are not recommended since it is assumed that each reagent will be tested to achieve optimal immunolabelling with minimal background interference.

6.2 Pre-embedding technique

Pre-embedding immunocytochemical techniques are used to identify antigenic sites on the outer membrane of isolated cells and micro-organisms (Fig. 2). This is one of the most sensitive techniques since fixation of the antigen is not an essential prerequisite of immunolabelling (Table 6.1). If very sensitive antigens

Table 6.1. *The pre-embedding technique*

1. Lightly fix cells with correct aldehyde (Chapter 5) or use unfixed.*
2. Incubate cells with:
   a. 1 per cent gelatin in phosphate-buffered (0.01 M, pH 7.2)–0.15 M NaCl (PBS) for 10 min.
   b. 0.02 M glycine in PBS 3 min.
   c. 1 per cent bovine serum albumin in PBS (PBS BSA) for 2 min.
   d. Antiserum diluted with PBS BSA (1 h).
   e. Wash 4 × 1 min with PBS BSA.
   f. Colloidal gold probe diluted with PBS BSA (1 h).
   g. Repeat e.
   h. Rinse with PBS (2 × 1 min).
3. Fix cells with 1 per cent glutaraldehyde in PBS at room temperature for 15 min.
4. Rinse briefly in water.
5. Post-fix with 1 per cent aqueous osmium tetroxide for 1 h.
6. Tertiary fix with 2 per cent aqueous uranyl acetate for 1 h.
7. Dehydrate in a graded series of ethanol.
8. Embed in Araldite or other preferred resin.
9. Cut ultrathin sections and stain with uranyl acetate and lead citrate before examination.

* If cells are not fixed, start immunolabelling schedule at 2 d.

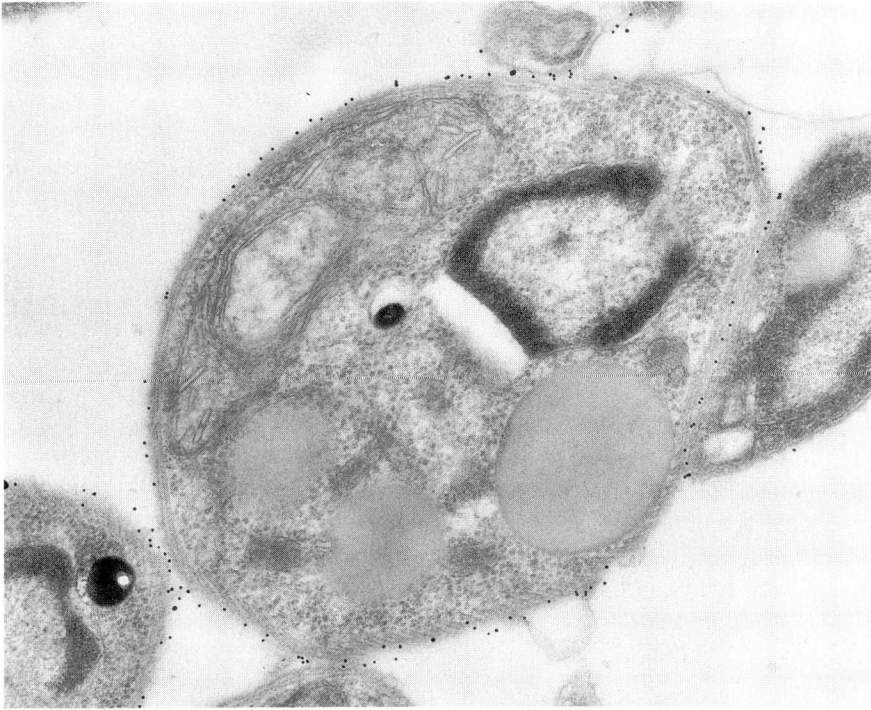

Fig. 2. The pre-embedding technique. Leishmania parasites were incubated with a biotinylated antibody raised against a surface protein. The gold probe in this case was therefore a streptavidin–gold complex. After immunolabelling, the preparation was fixed with glutaraldehyde, osmium tetroxide, and uranyl acetate before dehydration, embedding in epoxy resin, and sectioning. The sections, stained with uranium and lead salts, show immunolabelling associated with the surface of the organism. Magnification × 38 000. From a study in collaboration with Miss C. Scott, The Wellcome Research Laboratories, Beckenham, Kent, BR3 3BS, UK.

are being investigated, then labelling may be carried out at 4°C on unfixed tissue. A light aldehyde fixation is preferable, however, since this avoids any internalization of the reagents. After immunolabelling the sample is fixed with glutaraldehyde, osmium tetroxide, and uranyl acetate, then processed to ultrathin sections for electron microscope examination.

This technique produces good fine-structural preservation and excellent localization of external antigens. It has been used for localizing viral (Evans and Webb 1986) and bacterial (Mouton and Lamonde 1984; Orefici *et al.* 1986) antigens. Pre-embedding immunolabelling is used extensively to localize outer membrane antigens of peripheral blood cells (de Waele 1984), of Langerhans cell suspensions (Schmitt *et al.* 1984), of cultured cells such as lens fibre cells (Sas *et al.* 1985), and of human placental and cancer cells (Jemmerson *et al.* 1985). It is also a useful technique for studying antigens on isolated membranes (Kistler *et al.* 1985; Kordeli *et al.* 1986).

Localization of internal antigens necessitates permeabilizing the cells with detergents such as Triton X-100 to allow the reagents to penetrate the cells. This may cause antigen migration but (more seriously) false negative results might occur due to reagents not fully penetrating the cells. Localization of internal antigens using this technique is therefore not recommended for routine use, although it has been used successfully to localize filament-associated protein in baby hamster kidney cells (Yang *et al*. 1985) and also keratin filaments in human epidermal cells (Haftec *et al*. 1986).

An alternative method of allowing immunological reagents access to the interior of tissues, is to prepare vibratome sections of the tissue, which are immunolabelled and ultrathin resin sections are prepared for electron microscopic examination. Lamberts and Goldsmith (1985) have used this technique to advantage in the localization of β-endorphin-immunoreactive perikarya. It should be noted, however, that penetration of immunological reagents through a thick vibratome section is the same as for whole cell preparations. The cut surface layers only are labelled by the immunogold reagents, and therefore only the surface layers can be sectioned for electron microscopy.

6.3 Post-embedding technique

The post-embedding technique is the most widely used immunocytochemical technique for electron microscopic detection of antigens. The tissue is embedded and immunolabelling is carried out on sections of the tissue (Fig. 3). Any antigen exposed on the cut surface of the section is therefore available for immunolabelling. This technique is suitable for labelling both internal and external antigens, but the antigen must be able to withstand the processes necessary for obtaining sections of the tissue (Table 6.2).

Several embedding techniques are in routine use for post-embedding immunocytochemistry. Some workers prefer to dehydrate the sample with ethanol and then embed it in an epoxy resin such as Epon (Bendayan and Stephens 1984) or Araldite (Van Noorden and Polak 1985); others (Cramer *et al*. 1986) prefer methacrylate embedding, the tissue being dehydrated in a graded series of methacrylate. The acrylic resin LR White is gaining popularity (Yoshimura *et al*. 1986), the tissue for embedment being dehydrated totally in ethanol or in a mixture of ethanol and the resin itself. Other techniques, such as embedding in polyethylene glycol (Wolosewick *et al*. 1983) and embedding in epoxy resin after freeze-drying (Dudek *et al*. 1984), are not at present widely used, but may have application in the future.

Low-temperature embedding of the sample in the acrylic resin K4M is now a routine method (Carlemalm *et al*. 1982). It is believed that the progressive lowering of the temperature to $-40°C$ during dehydration in alcohol renders tissue proteins insoluble. They are therefore unlikely to leach from the tissue and, further, molecular movement is hindered at this temperature, again preventing leaching of cell components. These authors also claim reduced non-specific

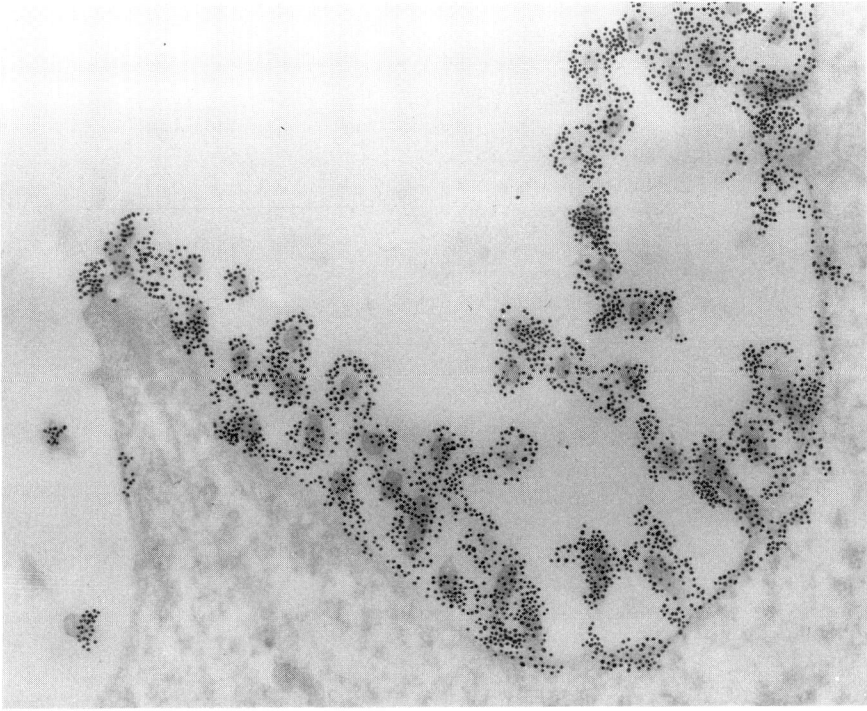

Fig. 3. The post-embedding technique using resin sections. Formaldehyde-fixed human breast tissue was dehydrated in alcohol and embedded in LR White resin. Sections of the tissue were incubated with mouse monoclonal antibody raised against epithelial cell antigen, rabbit anti-mouse Ig, and finally a 20 nm anti-rabbit Ig gold probe. The antigen is localized on the luminal surface of the ducts. Magnification × 25 500. From a study in collaboration with Dr S. Chantler and Dr U. Beckford, The Wellcome Research Laboratories, Beckenham, Kent, BR3 3BS, UK.

background labelling when using this resin. Despite the controversy concerning a possible temperature rise during polymerization (Ashford *et al.* 1986), this technique has produced many valuable results.

It has been suggested that the preparation of ultrathin frozen sections for immunolabelling is the method which yields optimum preservation of antigenic determinants, since the only potential denaturing step before immunolabelling is the initial aldehyde fixation (Griffiths *et al.* 1984). The tissue is aldehyde-fixed and, if a suspension, is embedded in 10 per cent gelatin which is subsequently hardened with the original fixative. Small pieces of sample are cryoprotected with 2.3 M sucrose for one hour to prevent ice crystal growth and concomitant tissue damage when the sample is subsequently plunged into either liquid nitrogen or liquid nitrogen slush. The sample is sectioned at − 80°C and blue cellophane-like sections are collected on a sucrose droplet. The sections are brought to room temperature and mounted on Butvar/carbon-coated grids for subsequent immunolabelling and staining (Tokuyasu 1984, 1986). The sections are stained

Table 6.2. *The post-embedding technique (after Slot and Geuze 1984)*

1. Prepare sections of fixed and either resin-embedded or frozen tissue on plastic and carbon-coated gold grid. Float these, sections downwards, on the following;
2. Phosphate-buffered (0.01 M, pH 7.2) saline (0.15 M) containing 1 per cent bovine serum albumin, PBS BSA (5 min).
3. Phosphate-buffered saline (PBS) containing 1 per cent gelatin (10 min).
4. PBS containing 0.02 M glycine (3 min).
5. Suitable dilution of antibody in PBS BSA (1 h).
6. Rinse 5 × 1 min in PBS BSA.
7. Suitable dilution of gold probe in PBS BSA (1 h).
8. Rinse (1 min) in PBS.
9. Fix with 1 per cent glutaraldehyde in PBS at room temperature for 3 min.
10. Rinse (5 × 1 min) in distilled water.
11. a. Resin sections
      i. Stain with uranyl acetate and lead citrate.
  b. Ultrathin frozen sections (Tokuyasu 1984)
      i. Float sections on 2 per cent neutral uranyl acetate† (10 min) to stabilize membranes.
     ii. Three rinses on droplets of water (20 s each).
    iii. Float sections on 2 per cent aqueous uranyl acetate (5 min).
    iv. Wash (3 × 20 s) on droplets of 1.5 per cent methyl cellulose (400 cps)‡ placed on a sheet of dental wax standing on ice.
     v. Pick up grid on a loop and remove sufficient methyl cellulose with the edge of a filter paper so that when dry a gold–blue interference colour remains.
    vi. Remove grid from loop and examine.

† 2 per cent neutral uranyl acetate is made by mixing equal volumes of aqueous solution of 4 per cent uranyl acetate and 0.3 M potassium oxalate and adjusting the pH to 7–8 by adding a small volume of 10 per cent ammonium hydroxide.
‡ A stock solution of methyl cellulose is made by suspending the required weight of powder in water at 60°C and then cooling to 4°C. It is diluted before use, the dilution being found experimentally to yield the correct thickness after drying the embedded sections.

with neutral and acidic uranyl acetate (Tokuyasu 1978) which produce a mixture of positive and negative contrast (Fig. 4). These sections possess no intrinsic rigidity and must be embedded in methyl cellulose before drying to maintain their three-dimensional integrity (Tokuyasu 1984, 1986).

With perhaps the exception of methacrylate, these embedding media are suitable for preparing and immunolabelling semithin sections for light microscopy to select an appropriate area for electron microscope analysis.

Whichever thin-sectioning technique is used, the sections mounted on plastic-coated gold grids are floated sequentially on droplets of the selected reagents (Table 6.3). During these incubations it is inadvisable to allow the grids to dry or else high levels of background labelling will be induced. Sections cut from frozen material should not be allowed to dry until after embedment in methyl cellulose. After cutting, these sections are held on droplets of phosphate-buffered saline containing 1 per cent bovine serum albumin until required for immunolabelling. Resin sections may be stored dry, for long periods, mounted on grids, awaiting immunolabelling.

The application of the post-embedding immunolabelling technique is so extensive that it has ramified into almost every biological discipline. Indeed, with

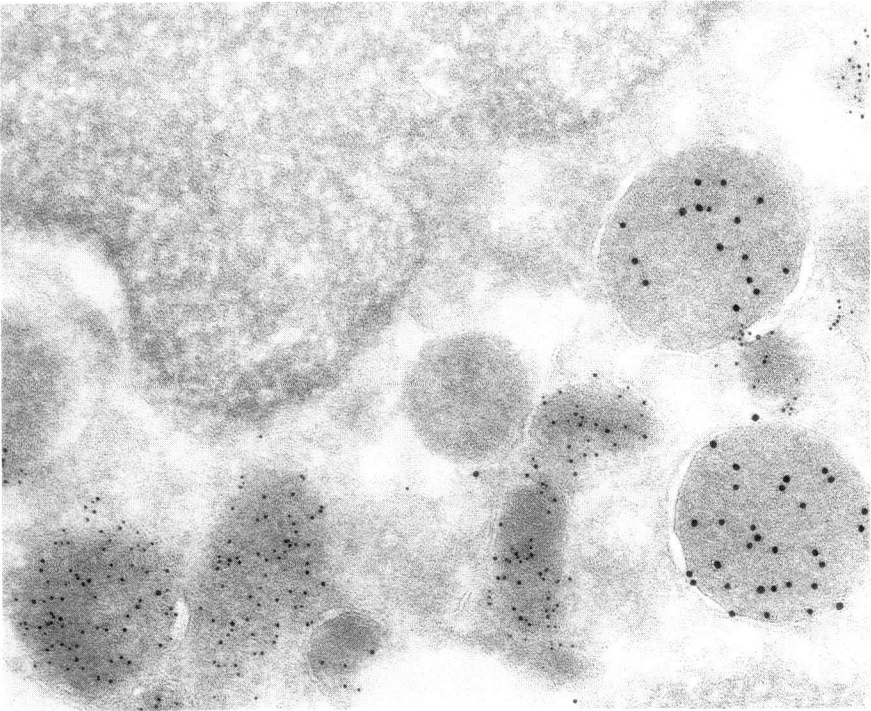

Fig. 4. The post-embedding technique using ultrathin frozen sections. Human granulocytes were formaldehyde-fixed and embedded in gelatin before cryoprotection in sucrose and freezing in liquid nitrogen slush. Ultrathin frozen sections were floated on a mixture of rabbit antibody raised against lactoferrin and mouse monoclonal antibody raised against elastase. After washing, the sections were incubated on a mixture of 15 nm gold spheres coated with goat anti-mouse Ig and 5 nm gold spheres coated with goat anti-rabbit Ig. The sections were stained with uranyl acetate. The double immunolabelling technique clearly shows elastase and lactoferrin in separate granules. Magnification × 8400. From a study in collaboration with Dr E. Cramer, Hopital Henri Mondor, 94010 Creteil, France. The monoclonal anti-elastase antibody was a gift from Dr D. Mason, John Radcliffe Hospital, Oxford, UK. Reproduced from J.E. Beesley (1987) *International Analyst*, 1, 20–5.

the exception of a few applications to specialized cell types, post-embedding techniques are almost universally employed for botanical electron immunocyto-chemistry using colloidal gold probes (Wang 1986), although theoretically any of the immunolabelling techniques described here could be used.

6.4 Scanning electron microscope technique

Horisberger *et al.* (1975) showed that colloidal gold probes were valuable for scanning electron microscope immunocytochemistry. It is only recently, however, that these techniques are being appreciated fully (Hodges *et al.* 1984, 1987; Park *et al.* 1986; Soligo *et al.* 1986; Studer and Herman 1986; Walther and Muller 1986). Immunoscanning electron microscopy is a very sensitive technique and possesses considerable potential for the study of external antigens (Fig. 5). It permits the examination of large pieces of tissue, thereby allowing a three-dimensional interpretation of the antigenic complex (Hodges *et al.* 1984).

The immunolabelling schedule is similar to the pre-embedding technique, but in this case the tissue, after immunolabelling, is prepared for scanning microscope observation (Table 6.3). Gold probes smaller than 20 mm diameter are not easily distinguished by use of the secondary electron imaging system so until recently the technique has been limited to using large 35 mm probes (Tetley *et al.* 1987).

Table 6.3. *The scanning electron microscope technique (after Hodges* et al. *1984)*

1. Select tissue according to the aims of the study and carry out pre-fixation in weak aldehyde (Chapter 5).
2. Wash specimens (3 × 5 min) with phosphate-buffered (0.01 M, pH 7.2) saline (0.15 M) containing 1 per cent bovine serum albumin (PBS BSA).
3. Incubate tissue with phosphate-buffered saline (PBS) containing 1 per cent gelatin (10 min).
4. Incubate tissue with PBS containing 0.02 M glycine (3 min).
5. Incubate tissue with primary antibody diluted with PBS BSA (1 h).
6. Wash specimens (3 × 5 min) with PBS BSA.
7. Incubate with gold probe diluted in PBS BSA (1 h).
8. Wash specimens (3 × 5 min) with PBS BSA.
9. Fix specimens in 2.5 per cent glutaraldehyde, buffered with 0.05 M sodium cacodylate pH 7.2.
10. Post-fix in 1 per cent aqueous osmium tetroxide (1–2 h) or process through the thiosemicarb-azide–osmium tetroxide schedule (Murphy 1980).
11. Dehydrate in a graded series of ethanol, ethanol, or acetone, critical point dry, deposit a thin conductive coating as necessary then examine using either secondary or backscattered electrons.

The technique has been refined by use of backscattered electrons for imaging the immunogold labelling (Soligo *et al.* 1986; de Harven *et al.* 1987). This imaging system relies on atomic contrast. Colloidal gold produces a much higher signal than the biological tissue and can be detected readily. Small (5–10 mm) probes may be used therefore to obtain maximal resolution. The technique could be used with extremely small gold probes in conjunction with the silver enhancement technique (Holgate *et al.* 1983), thereby increasing both resolution and minimum detection limit.

Hodges *et al.* (1984) are very enthusiastic with respect to the potential of this technique and feel that it offers a powerful approach to the study of cell membrane antigens. They believe that it is an attractive alternative to light microscope and transmission electron microscope examination of tissues, thereby providing a further spatial dimension in the analysis of surface membranes.

6.5 Negative stain technique

The immunonegative stain technique was developed specifically for the identification of external antigens on bacterial pili (Beesley *et al.* 1984*b*) and viruses (Beesley and Betts 1985). It is the technique of choice for immunolabelling

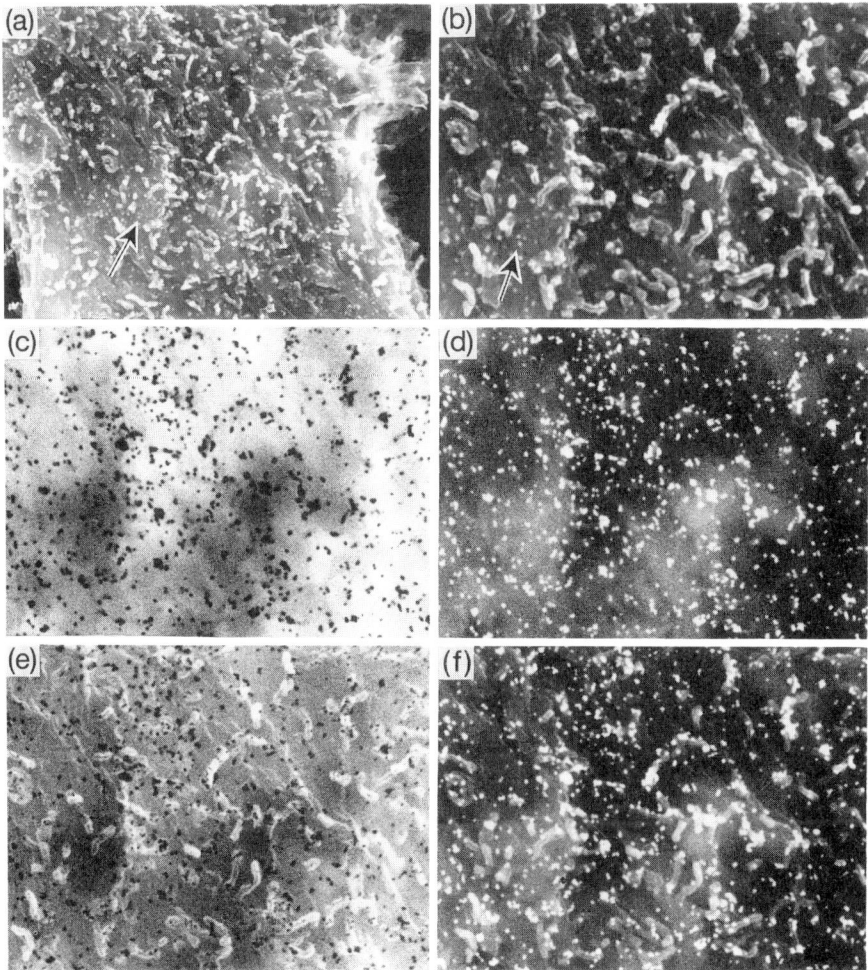

Fig. 5. Scanning electron microscope technique. A multiple display illustrating human colon-derived Ht29 cell surface expression of the L19 reactive epithelial membrane antigen marked by indirect labelling with 30 nm gold particles and imaged by scanning electron microscopy (SEM) using secondary (SE), backscattered (BE), or mixed (SE + BE) signals. Kindly provided by Dr G. Hodges, Imperial Cancer Research Fund, Lincoln's Inn Fields, London, WC2A 3PX, UK. Overview (a) and detail of cell surface (b) in the secondary electron imaging mode: limited number of gold particles (↑) detected (cf c–f). (c–f) The same field as seen in (b) but viewed: (i) In the backscattered electron imaging mode: the number of gold particles seen in the BE image is considerably greater than that in the SE image. BE image with reversed signal polarity when the gold particles appear as black dots (c); and normal signal polarity when the gold particles appear as white dots (d); (ii) By mixing the SE and BE signals a combined image is obtained in which surface morphology and gold labelling are simultaneously demonstrated; reversed BE signal polarity (e); and normal BE signal polarity (f). Magnification (a) × 6000, (b–f) × 12 000.

small particles which may be dried down on to a grid, immunolabelled *in situ*, then visualized by the negative stain method (Table 6.4).

This is a high-resolution technique. It is very quick, very simple, and it may be carried out using very little sample (Fig. 6). The gold probe produces immunolabelling, so distinctive that it may be detected even in samples

Table 6.4. *The negative stain technique (after Beesley* et al. *1984b)*

1. Dry sample on to Butvar/carbon-coated 400 mesh gold grid.
2. Float on droplet of antiserum diluted with phosphate-buffered (0.01 M, pH 7.2) saline (0.15 M) containing 1 per cent bovine serum albumin, PBS BSA (15 min).
3. Wash (4 × 1 min) by floating on droplets of PBS BSA.
4. Float on gold probe diluted with PBS BSA (15 min).
5. Wash by floating on droplets of water (4 × 1 min).
6. Negative stain, e.g. ammonium molybdate 1 per cent, pH 6.8.

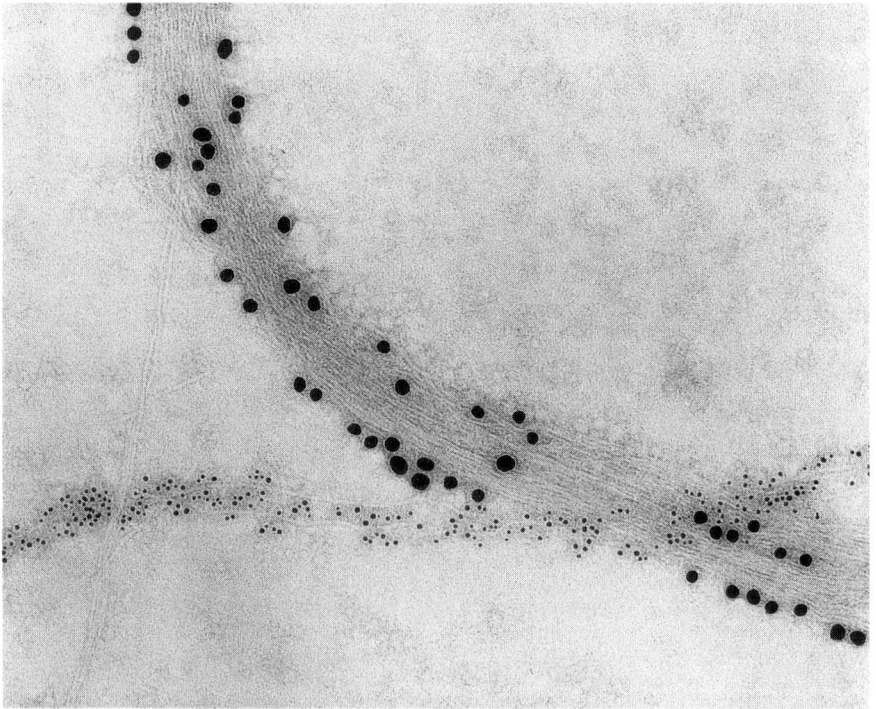

Fig. 6. The negative stain technique. Bacterial pili were dried on to a Butvar/carbon-coated 400 mesh gold grid and immunolabelled *in situ*. The preparation was sequentially incubated with the first antiserum, the 5 nm gold probe, the second antiserum, and then the 15 nm gold probe before negative staining. Three pili serotypes are observed; one detected with the first antibody and 5 nm probe, another detected with the second antibody and 15 nm probe, and the third serotype detected by its lack of reaction with either of the antibodies. Magnification × 91 000. Reproduced from J.E. Beesley (1987) *Serodiagnosis and Immunotherapy,* **1**, 239–52.

containing considerable debris. It is critical to select the correct antiserum for this technique since only external antigens of the sample are exposed. Antibodies directed against internal antigens do not immunolabel the sample.

6.6 Replica technique

The immunoreplica technique is a high-resolution technique for the localization of antigens on cultured cells (Fig. 7). The use of the colloidal gold techniques is of paramount importance in these studies since gold probes are probably the only immunocytochemical probes in routine use which can be identified easily on carbon–platinum replicas.

The immunoreplica technique was pioneered by Mannweiler *et al.* (1982) who used the protein A–gold technique to detect surface antigens on measles virus-infected cells. The infected tissue culture cells were fixed briefly then incubated with antibody and probe before preparing a replica of the immunolabelled cells (Table 6.5). Mannweiler *et al.* (1982) claim that immunolabelled sites can be recognized with the transmission electron microscope at magnifications as low as × 12 000. The gold probe is so small and so sensitive that the characteristic alterations of the plasma membrane induced by infection with the virus can be immunolabelled and still be detected. Colloidal gold probes are so dense that they may be visualized anywhere on the replica. This is an essential requirement for quantitative studies. The technique is ideal for searching relatively large areas of cell membrane for isolated patches of antigen.

Table 6.5. *The replica technique (after Mannweiler* et al. *1982)*

1. Cells cultured on coverslips are lightly fixed with aldehyde (Chapter 5).
2. Incubate (1 h) with antibody diluted with phosphate-buffered (0.01 M, pH 7.2) saline (0.15 M) containing 1 per cent bovine serum albumin (PBS BSA).
3. Wash (4 × 1 min) by incubating with PBS BSA.
4. Incubate (1 h) with gold probe diluted with PBS BSA.
5. Wash (4 × 5 min) with PBS BSA.
6. Wash with a suitable EM buffer such as 0.05 M sodium cacodylate pH 7.2 then post-fix with 2 per cent aqueous osmium tetroxide.
7. Dehydrate with ethanol.
8. Critical point dry.
9. Replicate with carbon and platinum before examination.

6.7 Freeze–fracture technique

There has been a growing interest in freeze–fracture cytochemistry using colloidal gold probes (Pinto da Silva 1984; Hohenberg *et al.* 1985; Berbel *et al.* 1986; Pinto da Silva *et al.* 1986). The freeze–fracture cytochemical techniques can be categorized as fracture–label (Pinto da Silva 1984) and label–fracture (Pinto da Silva and Kan 1984).

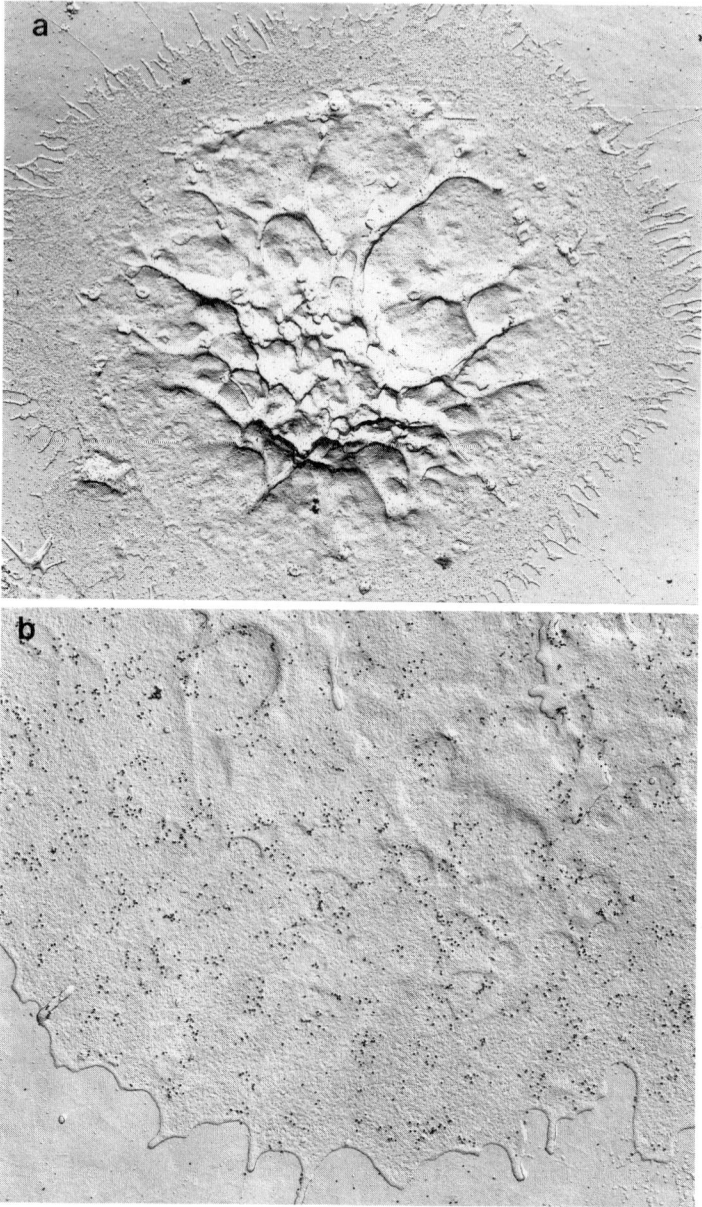

Fig. 7. The replica technique. Distribution of bovine serum albumin binding sites on cultured mouse peritoneal macrophages. The cells were incubated with bovine serum albumin–gold complexes for 1 hour at 4°C before dehydration, critical point drying, and replication. Survey micrograph (a) magnification 2400 and (b) higher power to show distribution of labelling on cell surface. Magnification 6800. Kindly donated by Priv-Doz Dr H. Robenek, Westfalische Wilhelms-Universitat, Munster, FRG.

Fracture–label can be further divided. Thin-section fracture–label (Table 6.6) is a high-resolution technique in which labelled fractured membranes can be examined in thin sections and compared with the overall fine structure of the tissue. Critical point drying fracture–label (Table 6.7) on the other hand produces a replica of the labelled surface of the fractured tissue. These replicas

Table 6.6. *Thin-section fracture–label technique for the detection of concanavalin A binding sites (after Pinto da Silva* et al. *1986)*

1. Fix cells with 1 per cent glutaraldehyde in phosphate-buffered isotonic saline, pH 7.4, 30 min at 4°C.
2. Embed (if a cell suspension) in 15 or 30 per cent bovine serum albumin at 25°C and gel with 1 per cent glutaraldehyde for 30 min at 25°C.
3. Slice gel into $1 \times 2 \times 2$ mm pieces and impregnate gradually with 30 per cent glycerol (1 h).
4. Freeze in Freon 22 cooled by liquid N_2.
5. Place the gels in a glass container (e.g. the base of a tissue homogenizer) filled with liquid nitrogen and cooled in a slush of liquid nitrogen and solid carbon dioxide.
6. Add drops of a 30 per cent glycerol, 1 per cent glutaraldehyde solution in 310 mOsmol phosphate buffer pH 7.5, in an amount approximately equal to that of the gels or tissue. Allow this and the gels to sediment.
7. Fracture gels with a glass pestle at $-190°C$ until pulverized into fine fragments.
8. Remove glass container to room temperature and allow nitrogen to reduce to 1/10 original.
9. Add 2 to 3 ml of glycerol–glutaraldehyde in buffer in liquid form.
10. Immerse glass container briefly in water bath at 30°C to thaw glycerol.
11. Transfer to an ice bucket for 15 min.
12. Deglycerinate and quench aldehyde groups by dropwise addition of 310 mOsmol sodium phosphate buffer pH 7.5 containing 1 mM glycyl-glycine.
13. Wash twice with sodium phosphate buffer.
14. Incubate fractured tissues or gels with 250 μg ml^{-1} concanavalin A in phosphate-buffered (0.01 M, pH 7.4) saline (0.15 M) containing 0.5 M $CaCl_2$ for 30 min at 25°C.
15. Wash in phosphate-buffered saline (PBS).
16. Incubate overnight at 4°C with colloidal gold complexed with horseradish peroxidase.
17. Wash in PBS.
18. Post-fix in buffered 1 per cent osmium tetroxide, 2 h at 4°C.
19. Tertiary fix with 0.5 mg ml^{-1} uranyl acetate in veronal acetate buffer pH 6.0, 90 min at room temperature.
20. Dehydrate in a graded series of ethanol or acetone and embed in resin of choice.
21. Cut thick sections to determine required area for electron microscope observation.
22. Cut ultrathin sections, stain with uranyl acetate and lead citrate, and examine.

Table 6.7. *Critical point drying fracture–label (after Pinto da Silva* et al. *1981, 1986)*

1. Prepare cells to Step 4, Table 6.6.
2. Transfer gels or tissues to a petri dish filled with liquid N_2 placed on top of liquid N_2/solid carbon dioxide slush.
3. Fracture specimen with liquid N_2-cooled scalpel.
4. Carry out Steps 8 to 18 of Table 6.6.
5. Dehydrate in ethanol and critical point dry with ethanol/carbon dioxide.
6. Attach gels, fracture side uppermost with double-sided sticky tape to specimen carrier and shadow with platinum and carbon and reinforce replica with carbon film.
7. Digest tissue in sodium hypochlorite.
8. Wash replicas with distilled water, mount on Formvar-coated grids and view.

Fig. 8. The freeze–fracture technique. Label–fracture of boar spermatozoa. Boar spermatozoa were labelled with wheat germ agglutinin and gold probes and then freeze–fractured. After thawing, the replicas were not treated with bleach or acids, but were washed with distilled water only. The label–fracture image provides a superimposition of the high-resolution conventional image of the exoplasmic fracture face and the high-resolution surface distribution of the label attached to receptors at the cell surface. Wheat germ agglutinin labelling is seen over the acrosomal region decreasing toward the neck. Over the particle-free zone at the base of the head, an unusually high concentration of receptors is observed. The striated cords below this zone are not labelled. Over the tail the receptors occur uniformly on both particle-free areas and intramembrane particle-rich areas. Magnification × 40 000. Reproduced from F.W.K. Kan and P.P. da Silva (1987) *Journal of Histochemistry and Cytochemistry*, **35**, 1069–78.

when viewed at low magnification are similar to orthodox freeze–fractured specimens, but at higher magnifications intramembranous paticles are not observed and the texture of membrane fracture faces suffers considerable alterations (Pinto da Silva 1984).

Label–fracture (Table 6.8) is a method for the high-resolution labelling of cell surfaces and permits the visualization of protein–gold complexes on cell surface sites over large cell surface areas. Pinto da Silva and Kan (1984) state that this

Table 6.8. *Label–fracture technique (Pinto da Silva and Kan 1984)*

1. Fix cells with 1.5 per cent glutaraldehyde in phosphate-buffered (0.01 M, pH 7.2)–0.15 M NaCl (PBS), 1 h at 4°C.
2. Wash twice in PBS.
3. Cytochemically label, e.g.:
   i. incubate with 0.25 mg ml^{-1} wheat germ agglutinin in PBS for 1 h at 37°C;
   ii. wash twice in PBS;
   iii. incubate with ovomucoid gold-complex diluted in PBS for 3 h at room temperature.
4. Impregnate fixed, labelled cells in 30 per cent glycerol.
5. Mount on double replica copper discs, freeze, freeze–fracture at −130°C, and replicate by Pt/C evaporation by conventional methods.
6. Wash replicas, at least six times, in distilled water, 30 min per wash.
7. Mount replicas on Formvar-coated grids for examination.

distribution is seen superimposed on the unaltered image of a freeze–fractured exoplasmic fracture face. The technique can be summarized as freeze–fracturing after labelling of cell surface receptors, but in contrast to conventional freeze–fracture, platinum–carbon replicas are not digested with acids or bases. They are, instead, repeatedly washed in distilled water and therefore the labelled outer half of the membrane remains attached to the replica. This permits the simultaneous observation of the surface label and the replica of the exoplasmic half of the plasma membrane.

Freeze-fracture labelling techniques appear to possess considerable potential for cytochemical labelling (Fig. 8). It is possible to label all membranes, both plasma and intracellular whether on isolated cells or within tissues as well as cytoplasmic components exposed by cross-fracture (Pinto da Silva 1984).

Colloidal gold immunolabelling techniques for light microscopy

7.1 Introduction

Light microscope immunocytochemistry remains the domain of the immuno-enzyme techniques. Colloidal gold techniques are, however, gaining popularity and it remains to be seen how these two techniques will coexist.

7.2 Immunolabelling isolated cells

The colloidal gold technique is now becoming increasingly popular for the study of surface antigens of peripheral blood cells (Table 7.1). Immunolabelling on these cells is readily visualized. The relatively small 16 nm gold particles can be recognized as a red ring around the cells when using bright field illumination. Gold particles are ideal for dark field observation (de Waele *et al.* 1983) when they appear as extremely bright spots on a dark background. Recently epipolarization microscopy has been used to examine the reaction product (de Waele *et al.* 1986), in which case the gold particles appear, as in dark field, as brilliant silver-grey spots but the advantage of the epipolarization technique is

Table 7.1. *Immunolabelling isolated human blood cells (de Waele* et al. *1983)*

1. Prepare mononuclear cells suspensions by Ficol-hypaque density gradient centrifugation.
2. Wash cells (3 × 5 min) in phosphate-buffered (0.01 M, pH 7.4) saline (0.15 M) containing 1 per cent bovine serum albumin, PBS BSA, and 1 per cent heat-inactivated normal AB serum (wash buffer).
3. Centrifuge and resuspend to 30×10^9 cells/litre in PBS BSA containing 4 per cent AB serum (incubation buffer).
4. Add 25 μl of diluted antibody (monoclonal raised in mouse) to 25 μl of cell suspension both suspended in incubation buffer.
5. Incubate at room temperature for 30 min.
6. Wash cells (3 × 5 min) with wash buffer.
7.* Resuspend cells in 25 μl of incubation buffer and 25 μl of goat anti-mouse IgG 40 nm gold probe.
8. Incubate for 1 h at room temperature.
9. Wash cells (3 × 5 min) with wash buffer.
10. Fix cells with 0.01 per cent glutaraldehyde in PBS pH 7.4 at room temperature for 10 min.
11. Prepare cytocentrifuge preparation.
12. Counterstain with 1 per cent chloroform-extracted aqueous methyl green.
13. Dehydrate and mount in synthetic mountant before examination.

* If silver enhancement technique is required carry on with Step 10, Table 7.2.

that the cell morphology can be visualized simultaneously with transmitted light. De Waele *et al.* (1986) have also applied the silver enhancement technique (Section 7.3) to the study of outer membrane antigens (Fig. 9). Colloidal gold immunolabelling of isolated cells appears to possess potential in clinical applications, and recent developments indicate a promising future in flow cytometric analysis (Bohmer and King 1984).

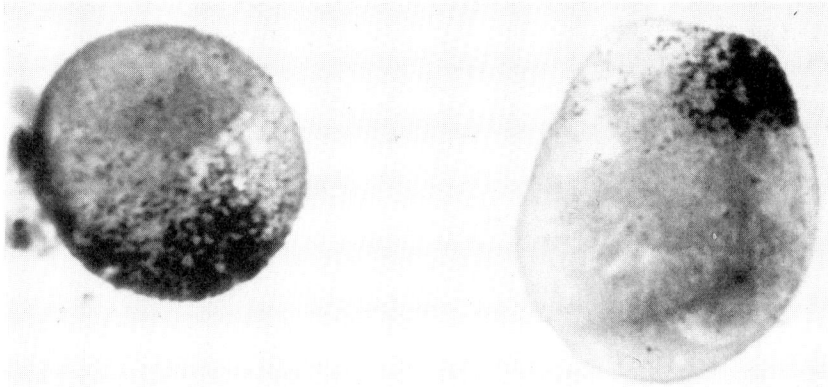

Fig. 9. Light microscope immunolabelling of isolated cells using the silver enhancement technique. Bone marrow aspirate stained with VIM 2 monoclonal antibody and Auroprobe LM GAM IgM followed by Intense II silver enhancement (Janssen Pharmaceutica, Beerse, Belgium). Bright field observation shows that VIM 2-positive cells have small dark granules on the surface membranes. Optimal morphological preservation is revealed by May–Grunwald counterstain. Magnification × 3800. Kindly provided by Dr M. de Waele, Vrije Universiteit, Brussels.

7.3 Post-embedding immunolabelling

Immunolabelling sections of tissue, the post-embedding technique, is extremely popular for light microscope examination of tissues. The technique is suitable for immunolabelling both internal and external antigens of cells but the antigen must be able to withstand the processes necessary for obtaining sections of tissue. The tissue may be embedded in either resin or wax for sectioning. Both of these preparative techniques require the antigen to be fixed. Unfixed or pre-fixed cryostat sections may be used if the antigen is extremely delicate or a rapid result is required.

Although gold immunolabelling appears as a red coloration when viewed with bright field microscopy, this is not sufficient if heavy counterstains are used for morphological identification of the tissue. The silver enhancement technique (Danscher and Norgaard 1983; Holgate *et al.* 1983), which yields a dense silver precipitate as the reaction product, is excellent for light microscope immunocyto-chemistry and is therefore becoming extremely popular (Fig. 10).

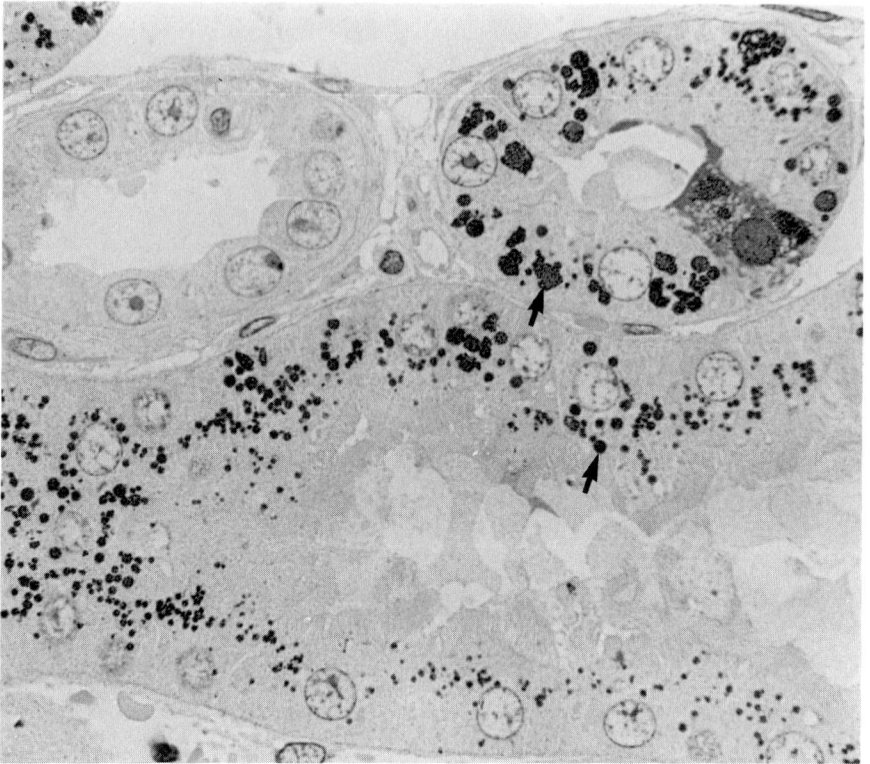

Fig. 10. Post-embedding immunogold–silver staining technique for light microscopy. Male rat kidneys were embedded in LR White and 1 μm sections were stained for the protein 'alpha$_{2u}$ globulin' with specific polyclonal antibody raised in rabbit and an immunogold–silver kit (Auro Probe LM Kit, Janssen Life Sciences Products, Oxford). The staining reaction (↑) is associated with the secondary lysosomes of the tubular cells. Magnification × 660. Kindly supplied by Dr N. Read, The Wellcome Research Laboratories, Beckenham, Kent, BR3 3BS, UK.

Sections of tissue, either resin, dewaxed, or frozen, are prepared and immunolabelled with antibody and the small (5 nm) gold probes to achieve the maximum number of probes on the section. In comparison with the larger (10–40 nm) gold probes, not only do these small probes possess more particles per unit volume, they actually penetrate the sections much more effectively and therefore give better labelling following silver enhancement (Lackie *et al.* 1985). The gold-labelled sections are then immersed in a silver solution and gold particles become nucleation centres on which reduced metallic silver is deposited; the particles grow and eventually become visible in transmitted light. This technique (Table 7.2) is extremely sensitive. It is quoted (Holgate *et al.* 1983) as being up to 200 times as sensitive as the immunoperoxidase technique.

Table 7.2. *The silver enhancement technique applied to sections of tissue (Holgate et al. 1983)*

1. Cut 5 μm paraffin sections of suitably fixed tissue.
2. Dewax and rehydrate.
   If sections of tissue have been embedded in epoxy resins, then the resin should first be removed (e.g. with sodium ethoxide–ethyl alcohol saturated with sodium hydroxide; Lane and Europa 1965), then the immunolabelling process begins here.
3. Apply Lugol's iodine for 5 min. (This appears to be a necessary step regardless of whether mercury has been included in the fixative.)
4. Rinse thoroughly and remove all traces of iodine with 2.5 per cent sodium thiosulphate.
5. If required, treat sections for 5 min with 0.1 per cent trypsin in Tris-buffered saline (0.05 M Tris in isotonic saline, pH 7.6) containing 0.1 per cent calcium chloride at 37°C.
6. Block non-specific background staining with normal serum from the species supplying the second antibody for 5 min and remove excess.
7. Treat sections with appropriately diluted primary antiserum in 5 per cent normal serum for 30 min.
8. Wash sections in Tris-buffered saline for 30 min.
9. Repeat Step 6.
10. Incubate with suitable immunogold reagent, preferably 5 nm, for 1 h.
11. Wash sections in Tris-buffered saline for 30 min, then distilled water for 30 min.
12.* Immerse sections in silver solution and develop under microscope control in darkroom (Safelight 5902 or F904).
13. Wash in distilled water. Fix briefly in 2.5 per cent sodium thiosulphate and wash in distilled water.
14. Counterstain with Mayers haematoxylin and eosin or any histological stain, dehydrate and mount (for wax sections). Resin sections can be counterstained with any routine stain before mounting in a proprietary medium or the resin itself.

---

* Recently, elegant silver enhancement kits have become commercially available (Janssen Pharmaceutica, Life Science Products, B2340 Beerse, Belgium). These usually involve mixing two solutions to generate the silver solution. This is applied to the section then followed with a second 'fixing' solution. These kits have been developed so that there is no need to carry out the process in the darkroom.

Alternatively, it is possible to prepare a silver solution in the laboratory. This consists of the following:

Gum acacia 500 g/l[a]	7.5 ml (diluted to 60 ml)
Citrate buffer, pH 3[b]	10 ml
Hydroquinone (0.85 g/15 ml)[c]	15 ml
Silver lactate (0.11 g/15 ml)[d]	15 ml

All solutions are made in distilled water and mixed in the above order.

a. Prepare by stirring overnight and filter through gauze. Stock solution can be frozen in aliquots.
b. 23.5 g trisodium citrate, $2H_2O$
   2.5 g citric acid, $1H_2O$
   100 ml distilled water
c. and d. Prepare just before use. Protect solutions containing silver lactate from light.

The silver enhancement technique is straightforward to perform and by virtue of its much enhanced sensitivity holds promise for the demonstration of tissue antigens.

Multiple labelling techniques

8.1 Introduction

The localization of one antigen within a tissue is valuable. To show the presence of two different antigenic sites can be invaluable. Colloidal gold probes are useful reagents for light microscope double labelling but their full potential for double labelling is exhibited in electron microscopy where individual particles of different sizes, localizing different binding sites can be seen.

Double labelling may be achieved using any of the techniques described in Chapters 6 and 7. Incubation times are not given, as the conditions stated in Chapters 6 and 7 will apply.

8.2 Light microscope techniques

Colloidal gold labelling appears as a red coloration in transmitted light, or a dark blue colour when viewed with phase contrast or Nomarski optics (Lucocq and Roth 1985). Alternatively, silver enhancement of the gold probes produces a black deposit. These colours may be combined with other colloidal metal markers or with immunoenzyme techniques which produce labelling of contrasting colours on the tissue.

Gu *et al.* (1981) favour the blue reaction product after revealing the immunoperoxidase label with 4-chloro-1-naphthol as the substrate, to contrast with the red of colloidal gold. Immunolabelling is carried out sequentially. The first antigen is labelled with the blue immunoperoxidase complex. If both the primary antibodies are raised in the same species, the immunocomplex from this reaction is removed by treating the sections with a glycine–hydrochloric acid solution [5 ml of 0.1 N HCl and 95 ml of glycine (0.75 per cent)–NaCl (0.58 per cent) mixture] for two hours. This is not necessary if the primary antibodies are raised in different hosts, such as mouse and rabbit.

The second antibody is revealed using a gold-labelled second antibody which produces the contrasting red coloration. If the acid treatment is carried out, that is if both primary antibodies are from the same host species, then it is necessary to carry out the immunoenzyme staining first. If acid treatment followed colloidal gold immunolabelling, the acid would remove the gold label attached to the antibody. In contrast, the enzyme reaction product is not attached to the antibody but is deposited in the tissue.

Roth (1982c) uses protein A complexed with colloidal silver and colloidal gold for immunolabelling two antigens in one preparation. The yellow of the colloidal silver probe contrasts with the red of the colloidal gold, each marking a different antigen. Immunolabelling is sequential and relies upon the first gold probe saturating all the available first antibody before the second antibody and probe are applied. In this manner, the first antibody is not immunolabelled with the second probe. It is expected, however, that the second antibody will bind to the first colloidal metal probe. The Fc portion of the antibody molecule will be bound to the protein A on the probe and would therefore not be available for further marking by the second probe. This technique is suitable for two antibodies raised either in identical or different host species.

Double labelling has been achieved using only one size colloidal gold probe (Manigley and Roth 1985). The first antigen was detected using specific antibody and the gold probe which was immediately converted into a black deposit by application of the silver enhancement technique. The second antibody was applied to the section and localized using the identical (14 nm) gold probe, without silver enhancement. This produced a red coloration in contrast with the black reaction product of the first label. Special care must be taken to use appropriately diluted reagents in the first incubation sequence, otherwise even traces of latent background (invisible gold probes) will become visible due to the silver enhancement process. Elegant double labelling has been achieved using this process. A further group (de Mey *et al.* 1986) have used the immunogold silver and peroxidase techniques in the same manner for double immunolabelling.

8.3 Electron microscope techniques
8.3.1 Introduction

Colloidal gold multiple labelling achieves supreme elegance in electron microscope studies since it is possible that, within a given sol, all the particles can be produced within a very narrow size range. The different size ranges are distinguished easily with the electron microscope, therefore it should be possible, theoretically, to label as many different reactive sites in tissue as there are different sized probes. In practice, however, experiments have generally been limited to using two different sized probes because factors such as cross-reactions and steric hindrance become extremely troublesome when using many reagents.

Several ingenious methods of electron microscope double labelling have been described.

8.3.2 Gold-labelled antibodies

The simplest technique for multiple labelling would be to coat gold spheres directly with primary antibody. Different antibodies could be complexed to different sized probes for multiple immunolabelling in a one-step procedure.

Providing the antibodies did not cross-react, multiple immunolabelling could be carried out simultaneously. This technique has not achieved widespread acceptance because to complex the antibody with the gold sol, relatively large volumes of antibody, which may be in short supply or be extremely expensive, are required. It has been suggested (Varndell and Polak 1984) that this problem will be alleviated in the future by the large-scale production of monoclonal antibodies. The technique has been used to identify high-density and low-density lipoprotein receptor sites on fibroblasts (Robenek and Severs 1984). In these experiments, the gold particles were complexed with either high- or low-density lipoprotein and used either sequentially or simultaneously for labelling cells before cell surface replicas were prepared.

Robinson *et al.* (1984) employ a variant of this technique to advantage for immunolabelling gonococcal macromolecules. They suspend protein A–gold complexes in specific antibody for 30 min at 25°C with agitation to produce the probe. Different sized probes are produced and double immunolabelling is carried out sequentially by immunolabelling with the first antibody/protein A–gold complex followed by the second antibody/protein A–gold complex.

8.3.3 Gold-labelled antigen detection

This technique was proposed by Larsson (1979). The pure antigen, which is used to raise antibody, is itself complexed with colloidal gold. Tissue-bound antigen is incubated with excess antibody so that one Fab site of the antibody remains free to bind with the antigen–gold complex. Both Fab sites on an antibody molecule are identical and therefore will bind with only one specific receptor. When used for multiple labelling experiments there is no risk of any cross-reaction occurring.

Unfortunately, this technique is not widely applied because highly purified or synthetic antigens are usually available only in small quantities either because of the tiny fraction remaining after purification or because they are prohibitively costly (frequently both). Furthermore, most of the antigen is generally needed to produce the antisera (Varndell and Polak 1984). This technique might be considered in extensive studies of single antigens.

8.3.4 Multiple antibody technique

The most elegant and easily applied technique for double labelling has been described by Tapia *et al.* (1983). Specific antibodies, raised in different hosts, are simultaneously applied to the tissue followed, after appropriate washes, by a suspension of two sizes of probes, each coated with an antibody raised against one of the two primary antibody species IgG and not cross-reacting with the other species. This is a very quick technique and cross-reactions between reagents are minimal (Fig. 4).

Unfortunately the necessary antibodies raised in different hosts are not always available.

8.3.5 Protein A-blocking technique

The protein A–gold technique has been successfully employed for labelling multiple antigenic sites in tissue sections (Roth 1982*b*; Slot and Geuze 1984), although in some cases high contamination rates have been reported (Varndell and Polak 1984). The principle of the technique is to label the first antigen with its specific antibody and a small (4 nm) protein A–gold probe. If free protein A is added (1 mg ml^{-1}), before the second incubation with second antibody and larger gold probe, it will saturate any free Fc sites on the preparation and should, therefore, block any cross-reactions by the second protein A–gold probe. Roth (1982*b*) and Slot and Geuze (1984) recommend that the first antigen should be localized with the smallest probe to minimize possible cross-reactions.

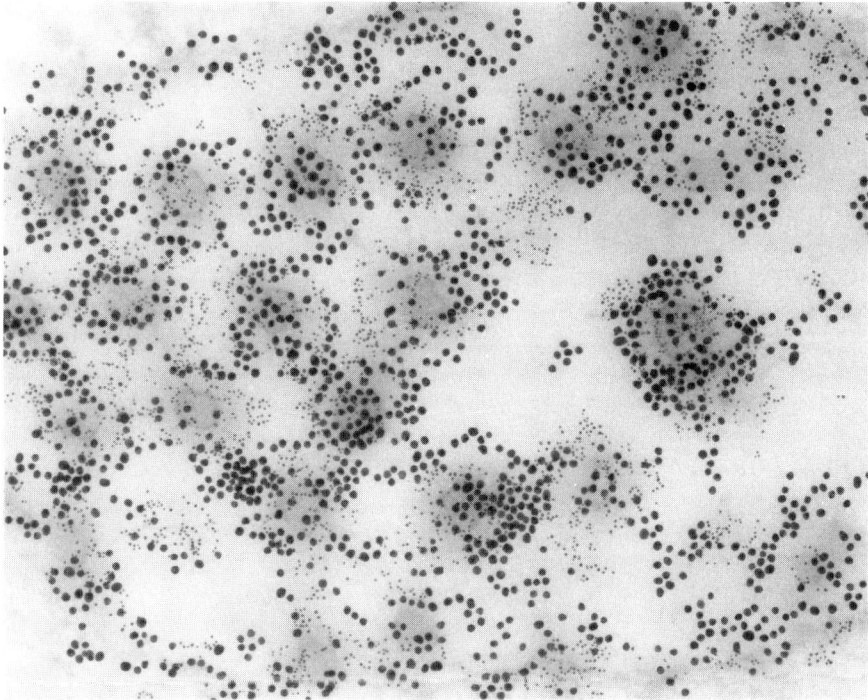

Fig. 11. Multiple immunolabelling resin sections using the double-sided labelling technique (Beesley *et al.* 1984). Human breast tissue was prepared and immunolabelled as described in Fig. 3. After drying the immunolabelling was coated with a thin layer of carbon to inhibit further labelling on this side of the section. A celloidin layer was deposited over this to strengthen the section. Immunolabelling with a different mouse monoclonal antibody was then carried out on the reverse side. Immunolabelling shows that both antigens are closely localized within the ductules. Magnification × 25 500. From a study in collaboration with Dr S. Chantler and Dr U. Beckford, The Wellcome Research Laboratories, Beckenham, Kent, BR3 3BS, UK.

8.3.6 Double-sided labelling technique

Bendayan and Stephens (1984) immunolabelled one side of the section with one antibody and a probe, then immunolabelled the reverse side with the second antibody and a second sized protein A–gold complex. The sections need to be mounted on uncoated grids. The application of this technique is limited therefore to resin sections which need no plastic support film.

Recent studies using immunogold probes have shown that despite careful manipulation, each face of the section can become contaminated with reagents applied to the opposite side (Beesley *et al.* 1984*a*). This problem is not insurmountable. One face of the section is immunolabelled with antibody and gold probe, then washed with water, dried in air, and carbon-coated. The thin layer of carbon prevents any further labelling on this surface of the grid. If desired, a celloidin layer may be deposited *after* the carbon coat to strengthen the section. Further reagents can then be applied to the opposite, uncoated face of the grid without fear of contaminating the reverse side (Fig. 11).

Controls

It is necessary that appropriate controls are carried out to monitor the immunological reactions. These tests should be a necessary extension of the controls for the efficiency of the antibodies (Section 2.4) and gold probes (Section 3.3.4).

Roth (1982a) recommends four controls for the protein A–gold reaction. These are:

1. Replacement of the primary antiserum with an antiserum previously incubated with its homologous antigen, followed by the application of the protein A–gold suspension.

2. Omission of the antiserum incubation step and application of the protein A–gold suspension alone.

3. Incubation of the thin sections with the specific antibody followed by a 1 h incubation with non-labelled protein A $(0.1–0.2 \text{ mg ml}^{-1})$ and then by the protein A–gold suspension.

4. Replacement of the antiserum with diluted non-immune serum of the species in which the antibody was raised followed by the protein A–gold suspension.

These, although described for the protein A–gold probe can be adapted to suit other types of probe. Naturally they should all produce either a negative result or a reasonably low background.

Control experiments for lectin cytochemistry involve pre-incubation of the binding sites with 0.1 M solution of the appropriate inhibitory sugar or with glycoprotein to the lectin or lectin–gold complex. The binding sites may also be incubated with an excess of native lectin $(1 \text{ mg ml}^{-1}$ for 30 min) before the application of the corresponding lectin–gold complex to which unlabelled lectin is also added. A further control is to treat the binding sites with 1 per cent periodic acid for 20 min before labelling. In two-step methods of labelling, the specificity of the reaction between lectin and glycoprotein–gold complex is checked by treatment of the tissue with 0.5 mg ml^{-1} of unlabelled glycoprotein between the two incubation steps (Roth 1983b).

Quantitation

Gold probes are ideal for quantitative electron microscopic studies because these dense, particulate markers can be recognized easily. The most satisfactory method applied so far to estimate the amount of gold labelling has been by counting the number of gold probes present on a given structure. Quantitative techniques have not yet been fully developed, nor are they used widely. For this reason, only an introductory review of colloidal gold quantitation is given.

It is absolutely necessary that quantitative studies are carefully controlled. Slight changes in fixation and embedding schedules probably cause subtle changes in the amount of labelling but so far these have not been examined in detail. Bendayan (1984a) maintains that quantitative evaluation should be carried out on sections specifically designated for quantitative experiments. The amount of specific labelling should be corrected to allow for the contribution of background labelling and he stresses that for comparative studies all tissue must be subject to identical fixation and embedding conditions and that all incubations within a given experiment must be carried out simultaneously.

The number of gold probes present on an immunolabelled structure is not a simple one-to-one ratio between probe and antigen. Gold probes larger than 3 nm possess more than one protein molecule, therefore gold probes larger than 3 nm are potentially capable of binding to more than one antibody. The 15 nm protein A–gold probe for instance bears about sixty protein molecules (Horisberger and Rosset 1977). A major advantage of using the small 2–3 nm probes for quantitative studies is that they possess just one protein molecule on each probe (Romano and Romano 1977). These should therefore approach a one-to-one probe-to-antigen ratio, but even this is doubtful where probes are in close apposition and effects of steric hindrance are possible.

Gold probes do not penetrate resin sections (Beesley and Adlam 1982; Bendayan 1984a). This facilitates quantitation using resin sections for post-embedding immunolabelling since variations in thickness of the section do not influence the amount of immunolabelling. Naturally for pre-embedding quantitative techniques, section thickness is a very important parameter.

Bendayan (1984a) suggests elegant techniques for quantitating the gold label. He uses point counting methods (Weibel 1969) or direct planimetry measurements to estimate the surface area of each structure evaluated for labelling density before the numbers of gold particles over each compartment are counted. Alternatively, a modular system for quantitative digital image analysis can be used for these evaluations. The density of labelling can then be estimated, and after taking background labelling into account, the true value calculated.

Bendayan and co-workers have quantitated the localization of pancreatic secretory proteins in subcellular compartments of the rat acinar cell (Bendayan *et al.* 1980) and also the concentration of amylase along its secretory pathway in the pancreatic acinar cell (Bendayan 1984*b*). Similar approaches have been applied to the localization of amelogenins, acidic glycoproteins of the enamel, revealed in the cellular compartments involved in protein secretion of the mouse ameloblasts as well as in the extracellular enamel organic matrix (Bendayan 1984*a*). Herbener *et al.* (1984, 1986) have also demonstrated the presence of gradients in the intensity of labelling for proteins from the rough endoplasmic reticulum to the Golgi complex and to the secretory granules which may reflect the protein concentration occurring along the rough endoplasmic reticulum–Golgi–granule secretory pathway.

These quantitative studies have compared the amount of label relative to a given experimental standard. Griffiths and Hoppeler (1986), however, have attempted to correlate the number of gold particles present on a section with the number of antigens present in that section. Their model system was baby hamster kidney cells infected with Semliki forest virus, and their antigens were well-characterized membrane protein complexes present in differing concentrations in the endoplasmic reticulum, Golgi stack, and virions present at the plasma membrane. They found that the efficiency of immunolabelling ultrathin frozen sections was 40, 13, and 12 per cent, respectively, for these three antigens. For Lowicryl K4M sections, however, these figures were reduced to 18.4, 6.6, and 1.2 per cent respectively. The authors conclude that the high efficiency of frozen sections was not entirely due to the complete penetration of reagents in the section, and the low efficiency of Lowicryl K4M sections was not entirely due to immunolabelling only the surface of the sections. Griffiths and Hoppeler (1986) point out that this study is only a first attempt to quantitate antigens and should be followed by more sophisticated studies.

In conclusion, although gold probes appear to possess considerable potential for quantitative studies, quantitation of colloidal gold labelling is still in its infancy. When the groundwork has been accomplished, however, quantitation of gold probes should be of immense value to the immunocytochemist.

Non-microscopical techniques

Overlay techniques for electrophoretically separated proteins transferred to immobilizing membranes are useful for the analysis of binding activities of proteins such as antibodies. One blot or blot unit is immunostained and the duplicate stained for overall protein pattern to correlate an immunodetected band and the total electropherogramme (Daneels *et al.* 1986). These techniques have proved to be important in molecular biology but, until recently, the practical value of these assays has been limited because of the complexity of the detection methods (Moeremans *et al.* 1984). Gold probes can be used as sensitive and specific detection methods for both electrophoretically-resolved and spot-blotted antigens and can alleviate the problems in detection (Fig. 12).

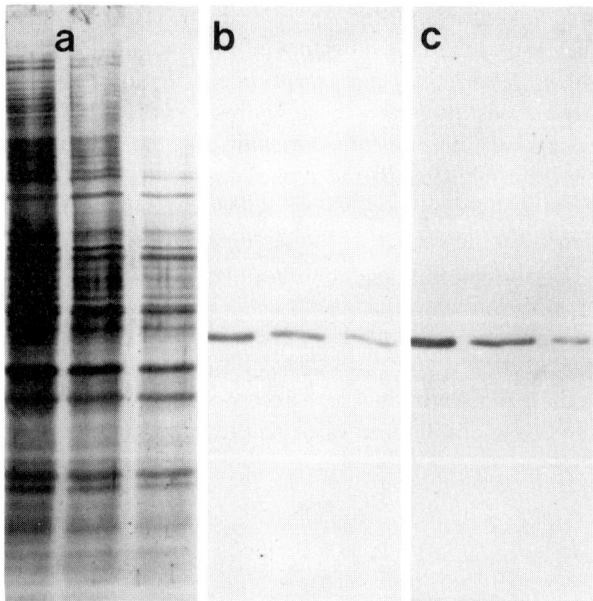

Fig. 12. Use of colloidal gold to detect both total protein staining (a) and for specific protein detection (b) and (c). (a) AuroDye forte staining of three twofold dilutions of a total cell extract of Chinese hamster ovary cells, separated by 7.5 per cent polyacrylamide gel (SDS-PAGE) and transferred to nitrocellulose. (b) Indirect Immunogold/Silver Staining (IGSS) of a duplicate blot reacted with a mouse monoclonal antibody directed against β-tubulin. (c) Streptavidin gold/silver staining of duplicate blot reacted with a mouse monoclonal antibody and subsequently with a biotinylated secondary antibody. Kindly donated by Dr M. Moeremans, Janssen Pharmaceutica, Beerse, Belgium.

Antigens are prepared by standard methods and are then immunolabelled with a suitable gold probe. The natural red colour of the attached probe is visible. The silver enhancement technique can be used for a more sensitive test (Brada and Roth 1984; Moeremans *et al*. 1984). This amplifies the gold signal so that weakly stained or previously absent bands become visible. With an antiserum against human growth hormone, Brada and Roth (1984) found a detection limit of 1 ng growth hormone with protein A–gold complex alone, and 0.1 ng growth hormone after the silver enhancement technique. This sensitivity is comparable to the autoradiographic methods of detecting proteins.

Comparison of immunogold and immunogold/silver staining methods with autoradiography and enzyme labelling methods demonstrates that colloidal gold probes are excellent markers for immune overlays (Moeremens *et al*. 1984). The gold technique has been shown to be specfic, and sensitive (Hsu 1984). Furthermore, the staining is rapid, there is no handling of radioactive materials, the stain is stable, there is no overstaining or understaining, and the procedure is inexpensive. No special equipment is needed so the testing can be performed in every laboratory (Brada and Roth 1984).

Problems commonly encountered during immunolabelling

12.1 Introduction

Colloidal gold immunolabelling has been in use for almost ten years and has many useful characteristics. The technique produces immunolabelling of scientific interest as well as aesthetic beauty, but as in all other techniques frustrating problems may arise. This chapter describes common practical problems which can and do occur during routine use of the probe.

Immunolabelling problems can be categorized under five headings: personal, no antigen, no antibody, no gold, and the wrong buffer. Lastly, of course, when these problems have been eliminated final trimming of the system may be carried out to extract the full potential of the technique. Naturally, not all of these problems are encountered in every sample.

12.2 Personal

The first category needs very little explanation. No one ever makes a mistake, but it has been known for grids to become muddled during handling of large numbers which must be kept in a strict sequence. They are then examined in the wrong order. 'Control' preparations then appear as the 'experimental' preparations and vice versa. Alternatively, one of the immunological reagents may have been omitted or used at the wrong dilution, or even the wrong reagent may have been used. For instance, the GAM (goat anti-mouse) serum may have been erroneously selected from the refrigerator instead of the GAR (goat anti-rabbit) serum. More seriously, instances can be found when the correct sample has been incorrectly processed or even the wrong sample has been processed. The remedy for these faults is, of course, obvious.

Finally, if immunolabelled structures are being examined with the electron microscope a brief check on the magnification used in relation to the size of the probe will inform the observer if it is actually possible to see the probes at the given microscope magnification.

12.3 No antigen

This may not be a fault of the system; it may be real. It is a difficult problem, especially for the novice who is more than ready to attribute the lack of antigen to faulty processing and immunolabelling of the tissue rather than to the tissue itself.

If no labelling is encountered after incubation with antibody and probe, valuable time will be saved if the antigen can be tested, preferably in the original state, by some other means. This will indicate whether it is the treatment which has destroyed the antigen or whether the sample, in the state presented for immunocytochemical analysis, is actually lacking antigen.

Fixation damage of the antigen is a constant problem. Slight fixation damage probably occurs all the while and passes unnoticed. Whenever fixation damage is noticed, it is usually a complete antigenic destruction and no immunolabelling results. The remedy is to transfer, if possible, to an alternative non-destructive fixative or change the technique to one which does not require fixation.

It is possible to use an artificial test substrate for evaluating the effects of fixation on the antigen (Gagne and Miller 1987). Polymerized bovine serum albumin blocks can be impregnated with antigen by soaking. The antigen may be used unfixed or treated as required. This system could be used to test not only for fixation artefacts but also for processing artefacts, it could be used also as a known positive system to check both the antibody (Section 2.4) and the protein–gold complexes (Section 3.3.4).

Antigen destruction by fixation should be contrasted with the effects of insufficient fixation. If there is insufficient fixation the antigen will leach out during processing giving rise to a redistribution of antigen rather than a negative result. This is a very real danger, when, in order to preserve more antigenicity lighter and lighter fixation regimes are used. There comes a point when unbeknown to the investigator the antigen will not be fixed properly, and leaches into surrounding tissues giving erroneous antigen localization.

Centrifugation is often a necessary procedure for embedding isolated organisms in resin. Some organisms, such as the bacterium *Pasteurella haemolytica*, contain an external loosely bound antigenic layer which is effectively removed by centrifugation (Beesley 1984). Therefore, organisms prepared by centrifugation may contain only a fraction of the original antigen.

Embedding the tissue in a resin may affect the antigen. Heat-labile antigens, for instance, may be destroyed by the heat needed to cure the resin. If this is suspected one of the 'low-temperature' resins such as Lowicryl K4M or cryoultramicrotomy should be used to prepare sections of tissue. At present there is some question as to the real temperatures encountered during polymerization of Lowicryl K4M (Ashford *et al.* 1986). Even when examining heat-stable antigens, different resins can affect the amount of immunolabelling obtained. Araldite sections, etched with hydrogen peroxide, produced very low levels of immunolabelling when compared with either methacrylate or cryosections (Walker and Beesley 1985). Van Noorden and Polak (1985), however, produce excellent immunolabelling using Araldite sections and Garzon *et al.* (1982) report immunolabelling Herpes virus particles in Epon sections. Antigens on structures smaller than the thickness of the section, such as viruses, may be masked by the resin in the section. The plane of the section may need to be known before a full interpretation of the immunolabelling is carried out, since the plane of section may appear to affect immunolabelling by masking certain antigenic sites.

12.4 No antibody. I. Physical inhibition of immunolabelling

No immunolabelling results if the antibody cannot get to the antigen. When immunolabelling electron microscope sections, be careful that the grids are floated specimen side down on the droplet of antiserum and that there are no air bubbles trapped under the grid to prevent the antibody from coming into contact with antigen. Also, when transferring grids from one droplet to another, be careful that the second reagent is not excessively diluted by liquid carried over from the first droplet.

Ultrathin frozen sections are particularly prone to folding. A corner of a section containing only supporting gelatin may fold back on the antigen, thus preventing immunolabelling. This may not be immediately apparent on high-magnification examination, since the resolution of the system is reasonable, even when viewed through a thin layer of gelatin. Low-magnification checks will confirm the presence of this artefact.

Finally, when studying viral antigens using the immunonegative stain technique it sometimes occurs that, although there may be ample antibody present in the serum, it is directed against an internal antigen which is masked by the specimen itself (Beesley and Betts 1984).

12.5 No antibody. II. The antiserum

This, as with no antigen (Section 12.3) may be real. The specific antibody may have been removed inadvertently from the serum during purification or the serum may have been raised against the wrong antigen.

Problems can occur when linking two antibodies. Always ensure that the second antibody will react with the first antibody in the experimental conditions employed. Remember also that protein A does not react similarly with all classes and subclasses of Ig from different species (Taatjes *et al.* 1987).

If the original antiserum was known to have contained antibody, but that activity has deteriorated check to determine whether the storage conditions were adequate. Never store diluted antibody in glass containers or a significant amount of the protein will be adsorbed on to the glass. Finally, always check the concentrations of antibodies. They are often used in highly dilute forms and if there has been any malfunction in the micropipettes used for measuring the antibody, the correct, very minute drop of antiserum will not have been transferred into the very large volume of buffer.

12.6 No gold probe

Techniques of gold probe production have been refined during the past few years, and excellent commercial probes are now available. Faulty gold probes are unlikely to occur unless the wrong probe has been used. If a protein A–gold probe has been used, a check on the affinity of protein A for the primary antibody and the pH at which the reaction is carried out is worthwhile. If it appears likely that the gold is at fault, check that the probe has not been denatured through incorrect storage and check that the probe has not been used at the wrong concentration. If the gold probe has been manufactured in the laboratory, check that the quantity of protein used to stabilize the gold sol is sufficient.

If the gold has been used for pre-embedding immunolabelling of small organisms, such as bacteria, check whether the primary antibody has caused the organisms to clump, thereby excluding the gold probe from the innermost organisms. A section including the outermost organisms of the pellet will determine whether or not this fault is present.

12.7 Faulty buffer system

Naturally the buffer must be at the correct pH or else the antigen–antibody complex will dissociate. This problem is unlikely to occur if standard buffer recipes are followed. A problem may occur if distilled water is used for washing immunolabelled preparations. The distilled water may be at a very low pH which will cause the antigen–antibody complex to dissociate, thereby reducing immunolabelling (Duerrenberger *et al.* 1986). Although the antigen–antibody reaction is reversible, prolonged washing has not been found to remove the high-affinity, specific antibodies attached to antigen (Beesley *et al.* 1983).

The pH of the buffer can affect the amount of immunolabelling obtained, especially if the protein A–gold system is used. Protein A, for instance, associates with human, rabbit, and guinea-pig antibody at neutral pH, but needs a higher pH to associate completely with mouse, rat, and goat antibody.

The wrong buffer components can also lead to high background immuno-labelling which may camouflage the specific labelling. Suitable buffer recipes, which contain at least one blocking agent to reduce high background labelling are given where appropriate with the descriptions of the various immunolabelling techniques.

12.8 Final adjustments

Once labelling has been obtained, visual examination of the results will indicate whether the tissue is over- or under-labelled. Final trimming of the system, such as varying the concentration of reagents or time of labelling will bring out the full potential of the system.

12.9 Conclusions

This chapter has been a record of practical, everyday problems in the use of the colloidal gold probes. As such it is intended to be used as a very practical guide to trouble-shooting if problems are experienced. This chapter is not intended to replace the application of carefully executed control experiments.

The future

Colloidal gold immunolabelling techniques are now well established among the scientific community as highly respectable techniques which have already produced a wealth of scientific knowledge. Colloidal gold has many positive attributes and this new revolution in cytochemical marking has gained tremendous momentum for widespread application in the biological sciences. It is highly likely that this widespread applicability will continue for many years. Despite their establishment, the colloidal gold techniques are by no means static and several new techniques are being developed which should be extremely useful in the future.

Recent developments in microinjection techniques have led to the exciting possibilities of injecting colloidal gold-labelled reagents into cells. After microinjection, the cells may be prepared for ultrastructural examination by fixation, dehydration, embedding in resin, and thin sectioning (Wehland and Willingham 1983; Schulze and Kirschner 1986). Another exciting possibility is to detect the gold particles, whilst the cells are still living, with a video recorder using transmitted light and electronic subtraction of diffuse background light. These techniques, called nanometre particle video ultramicroscopy, or nanovid ultramicroscopy (de Brabander et al. 1985, 1986) are already producing novel dimensions in the study of intracellular mobility, cell membrane dynamics, receptor translocation, internalization, and intracellular routing of proteins, and should prove of immense value in furthering our understanding of the living cell.

Injected colloidal gold has also been employed for localizing the centre of sites of the microinjection of drugs in the brain. Colloidal gold, in conjunction with the silver enhancement technique, provided a simple method, with a high degree of stability and spatial resolution, for the localization of the microinjection centres (Hodgson et al. 1987).

Photoelectron microscopy images of biological structures are maps of the ultraviolet-stimulated electron emission originating from the outer 10 nm or less of the cell surface. Image contrast arises from the effect of topography on the trajectories of emitted electrons and differences in the photoelectron quantum yields between surface components (Birrell et al. 1986). This is a useful imaging technique with which to observe colloidal gold probes on the surface of the cell. Birrell et al. (1986) have found that 6 nm colloidal gold particles in conjunction with the silver enhancement technique provide clear labelling patterns and an image of the uncoated sample background. It appears that this technique will become a useful complement to the immunoscanning and immunoreplica techniques.

Another exciting colloidal gold technique that has emerged recently is that of *in situ* hybridization at the electron microscope level of resolution (Binder *et al.* 1986). This technique, using sections of Lowicryl K4M embedded tissue, is still under development but eventually it should permit the direct investigation of gene expression and its modulation of the individual cell during its developmental transitions. This technique should soon be firmly established among the repertoire of colloidal gold techiques.

Finally, protein G, a cell wall protein isolated from human group G streptococci strain G148 has been proposed as a novel gold conjugate, possibly superior to the protein A–gold probe for immunocytochemistry (Bendayan 1987). There appears to be some controversy as to its superior nature (Taatjes *et al.* 1987) and the ensuing literature will be followed with interest.

In conclusion, colloidal gold techniques are dominating electron immunocyto-chemistry and are gaining in popularity for light microscopy. Colloidal gold techniques have many admirable characteristics for both research and diagnostic studies and have shown themselves to be highly versatile and highly adaptable, enabling them to keep pace with modern trends in biological microscopy. They should, therefore, remain as routine immunocytochemical techniques for many years into the future.

References

Ashford, A.E., Allaway, W.G., Gubler, F., Lennan, A., and A. Sleegers J. (1986). Temperature control in Lowicryl K4M and glycol methacrylate during polymerisation: is there a low temperature embedding method? *Journal of Microscopy*, **144**, 107–26.

Bao-le Wang, Scopsi, L., Hartvig-Nielson, M., and Larsson, L-I. (1985). Simplified purification and testing of colloidal gold probes. *Histochemistry*, **83**, 109–15.

Baschong, W., Lucocq, J.M., and Roth, J. (1985). 'Thiocyanate gold': small (2–3 nm) colloidal gold for affinity cytochemical labelling in electron microscopy. *Histochemistry*, **83**, 409–11.

Bauer, H., Farr, F.R., and Horisberger, M. (1974). Ultrastructural localisation of cell wall teichoic acids in *Streptococcus faecalis* by means of concanavalin A. *Archives of Microbiology*, **97**, 17–26.

Beesley, J.E. (1984). Recent advances in microbiological immunocytochemistry. In *Immunolabelling for electron microscopy* (ed. J.M. Polak and I.M. Varndell), pp. 289–303. Elsevier Science Publishers, BV, Amsterdam.

Beesley, J.E. (1987). Colloidal gold electron immunocytochemistry: its potential in medical microbiology. *Serodiagnosis and Immunotherapy*, **1**, 239–52.

Beesley, J.E. and Adlam, C. (1982). The protein A–gold technique: a comparison between the staining mechanism on methacrylate and cryosections. *Proc. 10th Int. Cong. E.M.*, **3**, 265–6.

Beesley, J.E. and Betts, M.P. (1984). Applications of the immunonegative stain technique to the study of viral and bacterial antigens. *Proc. 8th Eur. Cong. E.M.*, **3**, 1595–6.

Beesley, J.E. and Betts, M.P. (1985). Virus diagnosis: a novel use for the protein A–gold probe. *Medical Laboratory Sciences*, **42**, 161–5.

Beesley, J.E., Orpin, A., and Adlam, C. (1983). An evaluation of the conditions necessary for optimal protein A–gold labelling of capsular antigen in ultrathin methacrylate sections of the bacterium *Pasteurella haemolytica*. *Histochemical Journal*, **16**, 151–63.

Beesley, J.E., Beckford, U., and Chantler, S.M. (1984*a*). Double immunogold staining method for the simultaneous localisation of two tumour associated epithelial cell antigens in breast tissue. *Proc. 8th Eur. Cong. E.M.*, **3**, 1599–1600.

Beesley, J.E., Day, S.E.J., Betts, M.P., and Thorley, C.M. (1984*b*). Immunocytochemical labelling of *Bacteroides nodosus* pili using an immunogold technique. *Journal of General Microbiology*, **130**, 1481–7.

Behnke, O. *et al* (1986). Non-specific binding of protein-stabilised gold sols as a source of error in immunocytochemistry. *European Journal of Cell Biology*, **41**, 326–38.

Bendayan, M. (1984*a*). Protein A–gold electron microscopic immunocytochemistry: methods, applications and limitations. *Journal of Electron Microscope Technology*, **1**, 243–70.

Bendayan, M. (1984*b*). Concentration of amylase along its secretory pathway in the pancreatic acinar cell as revealed by high resolution immunocytochemistry. *Journal of Histochemistry*, **16**, 85–108.

Bendayan, M. (1985). The enzyme–gold technique: a new cytochemical approach for the ultrastructural localisation of macromolecules. In *Techniques in immunocytochemistry*, Vol. 3 (ed. G.R. Bullock and P. Petrusz), pp. 179–201. Academic Press, London.

Bendayan, M. (1987). Introduction of the protein G–gold complex for high resolution immunocytochemistry. *Journal of Electron Microscope Technology*, **6**, 7–13.

Bendayan, M. and Stephens, H. (1984). Double labelling cytochemistry applying the protein A–gold technique. In *Immunolabelling for electron microscopy* (ed. J.M. Polak and I.M. Varndell), pp. 143–54. Elsevier Science Publishers, BV, Amsterdam.

Bendayan, M. and Zollinger, M. (1982). Protein A–gold immunocytochemical techniques: ultrastructural localisation of antigens on post-osmicated tissues. *Proc. 10th Int. Cong. E.M.*, **3**, 267–8.

Bendayan, M., Roth, J., Perrelet, A., and Orci, L. (1980). Quantitative immunocytochemical localisation of pancreatic secretory proteins in subcellular compartments of the rat acinar cell. *Journal of Histochemistry and Cytochemistry*, **28**, 149–60.

Berbel, J.V., Steup, M., Volker, W., Robenek, H., and Flugge, U.I. (1986). Polypeptides of the chloroplast envelope membranes as visualised by immunochemical techniques. *Journal of Histochemistry and Cytochemistry*, **34**, 577–83.

Binder, M., Tourmente, S., Roth, J., Renaud, M., and Gehring, W.J. (1986). In situ hybridisation at the electron microscope level: localisation of transcripts on ultrathin sections of Lowicryl K4M embedded tissue using biotinylated probes and protein A–gold complexes. *Journal of Cell Biology*, **102**, 1646–53.

Birrell, G.B., Habliston, D.L., Hedberg, K.K., and Griffith, O.H. (1986). Silver enhanced colloidal gold as a cell surface marker for photoelectron microscopy. *Journal of Histochemistry and Cytochemistry*, **34**, 339–45.

Birrell, G.B., Hedberg, K.K., and Griffith, O.H. (1987). Pitfalls of immunogold labelling: analysis by light microscopy, transmission electron microscopy and photoelectron microscopy. *Journal of Histochemistry and Cytochemistry*, **35**, 843–53.

Bohmer, R.M. and King, N.J.C. (1984). Immunogold labelling for flow cytometric analysis. *Journal of Immunological Methods*, **74**, 49–57.

Brada, D. and Roth, J. (1984). 'Golden Blot' – detection of polyclonal and monoclonal antibodies bound to antigens on nitrocellulose by protein A–gold complex. *Analytical Biochemistry*, **142**, 79–83.

Carlemalm, E., Garavito, R.M., and Villiger, W. (1982). Resin development for electron microscopy and an analysis of embedding at low temperature. *Journal of Microscopy*, **126**, 123–43.

Cramer, E.M., Caen, J.P., Drouet, L., and Breton-Gorius, J. (1986). Absence of tubular structures and immunolabelling for von Willebrand factor in the platelet α-granules from porcine von Willebrand disease. *Blood*, **68**, 774–8.

Daneels, G., Moeremans, M., de Raemaeker, M., and de Mey, J. (1986). Sequential immunostaining (gold/silver) and complete protein staining (aurodye) on western blots. *Journal of Immunological Methods*, **89**, 89–91.

Danscher, G. and Norgaard, J.O.R. (1983). Light microscopic visualisation of colloidal gold on resin embedded tissue. *Journal of Histochemistry and Cytochemistry*, **31**, 1394–8.

De Brabander, M., Geuens, G., Nuydens, R., Moeremans, M., and De Mey, J. (1985). Probing microtubule-dependent intracellular mobility with nanometre particle video ultramicroscopy (nanovid ultramicroscopy). *Cytobios*, **43**, 273–83.

De Brabander, M., Nuydens, R., Geuens, G., Moeremans, M., and de Mey, J. (1986). The use of submicroscopic gold particles combined with video contrast enhancement as a simple molecular probe for the living cell. *Cell Mobility and the Cytoskeleton*, **6**, 105–13.

Debray, H., Decout, D., Strecker, G., Spik, G., and Montreuil, J. (1981). Specificity of twelve lectins towards oligosaccharides and glycopeptides related to *N*-glycosylproteins. *European Journal of Biochemistry*, **117**, 41–55.

De Harven, E., Soligo, D., and Christensen, H. (1987). Should we be counting gold marker particles on cell surfaces with the SEM? *Journal of Microscopy*, **146**, 183–9.

De Mey, J., Hacker, G. W., De Waele, M., and Spingall, D.R. (1986). Gold probes in light microscopy. In *Immunocytochemistry: modern methods and applications* (2nd edn) (ed. J.M. Polak and I.M. Varndell), pp. 71–88. J. Wright & Sons, Bristol.

De Waele, M. (1984). Haematological electron immunocytochemistry. In *Immunolabelling for electron microscopy* (ed. J.M. Polak and I.M. Varndell), pp. 267–88. Elsevier Science Publishers, BV, Amsterdam.

De Waele, M., de Mey, J., Moeremans, M., de Brabander, M., and Van Camp, B. (1983). Immunogold staining method for the light microscopic detection of leukocyte cell surface antigens with monoclonal antibodies. *Journal of Histochemistry and Cytochemistry*, **31**, 376–81.

De Waele, M., de Mey, J., Renmans, W., Labeur, C., Jochmans, K., and Van Camp, B. (1986). Potential of immunogold–silver staining for the study of leukocyte subpopulations as defined by monoclonal antibodies. *Journal of Histochemistry and Cytochemistry*, **34**, 1257–63.

Dudek, R.W., Varndell, I.M., and Polak, J.M. (1984). Combined quick-freeze and freezedrying techniques for improved electron immunocytochemistry. In *Immunolabelling for electron microscopy* (ed. J.M. Polak and I.M. Varndell), pp. 235–48. Elsevier Science Publishers, BV, Amsterdam.

Duerrenberger, M., Carlemalm, E., Villiger, W., and Kellenberger, E. (1986). Low temperature procedures and chemical coupling of antigen–antibody–protein A complex improve immunocytochemical labelling. *Proc. 11th Int. Cong. E.M.*, **3**, 2303–4.

Evans, N.R.S. and Webb, H.E. (1986). Immunoelectron-microscopical labelling of glycolipids in the envelope of a demyelinating brain-derived RNA virus (Semliki Forest) by anti-glycolipid sera. *Journal of Neurological Science*, **74**, 279–87.

Faulk, W.P. and Taylor, G.M. (1971). An immunocolloid method for the electron microscope. *Immunochemistry*, **8**, 1081–3.

Faulk, W.P., Vyas, G.N., Phillips, C.A., Fudenberg, H.H., and Chism, K. (1971). Passive haemagglutination test for anti-rhinovirus antibodies. *Nature*, **231**, 101–4.

Frens, G. (1973). Preparation of gold dispersions of varying particle size: controlled nucleation for the regulation of the particle size in monodisperse gold suspensions. *Nature: Physical Science*, **241**, 20–2.

Gagne, G.D. and Miller, M.F. (1987). An artificial test substrate for evaluating electron microscopic immunocytochemical labelling reactions. *Journal of Histochemistry and Cytochemistry*, **35**, 909–16.

Garzon, S., Bendayan, M., and Kurstak, E. (1982). Ultrastructural localisation of viral antigens using the protein A–gold technique. *Journal of Virological Methods*, **5**, 67–73.

Geoghegan, W.D. (1985). The adsorption of IgG and IgG fragments to colloidal gold: molecular orientation. *Journal of Cell Biology*, **101**, 85a.

Geoghegan, W.D. (1986). The adsorption of rabbit IgG to colloidal gold: molecular orientation. (Proc. Histochem. Soc. America.) *Journal of Cell Biology*, **32**, 1360.

Geoghegan, W.D. and Ackerman, G.A. (1977). Adsorption of horseradish peroxidase, ovomucoid and anti-immunoglobulin to colloidal gold for the indirect detection of concanavalin A, wheat germ agglutinin and goat anti-human immunoglobulin G on cell surfaces at the electron microscopic level: a new method, theory and application. *Journal of Histochemistry and Cytochemistry*, **25**, 1182–200.

Griffiths, G. and Hoppeler, H. (1986). Quantitation in immunocytochemistry: correlation of immunogold labelling to absolute number of membrane antigens. *Journal of Histochemistry and Cytochemistry*, **34**, 1389–98.

Griffiths, G., McDowall, A., Back, R., and Dubochet, J. (1984). On the preparation of cryosections for immunocytochemistry. *Journal of Ultrastructure Research*, **89**, 65–78.

Gu, J., De Mey, J., Moeremans, M., and Polak, J.M. (1981). Sequential use of the PAP and immunogold staining method for the light microscopical double staining of tissue antigens. *Regulatory Peptides*, **1**, 365–74.

Haftec, M., Staquet, M.J., Viac, J., Schmitt, D., and Thivolet, J. (1986). Immunogold labelling of keratin filaments in normal epidermal cells with two anti keratin monoclonal antibodies. *Journal of Histochemistry and Cytochemistry*, **34**, 613–18.

Handley, D.A. and Chien, S. (1987). Colloidal gold labelling studies related to vascular and endothelial function, hemostasis and receptor-mediated processing of plasma macromolecules. *European Journal of Cell Biology*, **43**, 163–74.

Herbener, G.H., Bendayan, M., and Feldhoft, R.C. (1984). The intracellular pathway of vitellogenin secretion in the frog hepatocyte as revealed by protein A–gold immuno-cytochemistry. *Journal of Histochemistry and Cytochemistry*, **32**, 697–704.

Herbener, G.H., Bendayan, M., and Feldhoft, R.C. (1986). Albumin localization in the frog hepatocyte during vitellogenesis as revealed by protein A–gold immunocyto-chemistry. *Journal of Histochemistry and Cytochemistry*, **34**, 665–71.

Hodges, G.M., Smolira, M.A., and Livingston, D.G. (1984). Scanning electron microscope immunocytochemistry in practice. In *Immunolabelling for electron micro-scopy* (ed. J.M. Polak and I.M. Varndell), pp. 189–233. Elsevier Science Publishers, BV, Amsterdam.

Hodges, G.M., Southgate, J., and Toulson, E.C. (1987). Colloidal gold—a powerful tool in S.E.M. immunocytochemistry: an overview of bioapplications. *Scanning Micro-scopy*, **1**, 301–18.

Hodgson, A., Minson, J., Chalmers, J., and Pilowsky, P. (1987). The use of microinjected colloidal gold and immunocytochemistry to localise pressor sites in the rostral medulla oblongata of the rat. *Neuroscience Letters*, **77**, 125–30.

Hohenberg, H., Rutter, G., Bohn, W., and Mannweiler, K. (1985). Direkts korrelation zwischen Antigenen und Ultrastrukturen aut verschiedenen. Ebenen der Plasmamem-bran mit Hilfe der Immunogold-und Replikatechnik. *Beitrage zur Elektronenmikrosko-pischen Direktabbildung von Oberflachen*, **18**, 313–22.

Holgate, C.S., Jackson, P., Cowen, P.N., and Bird, C.C. (1983). Immunogold silver staining: a new method of immunostaining with enhanced sensitivity. *Journal of Histochemistry and Cytochemistry*, **31**, 938–44.

Horisberger, M. (1985). The gold method as applied to lectin cytochemistry in transmission and scanning electron microscopy. In *Techniques in immunocytochemistry*, Vol. 3 (ed. G.R. Bullock and P. Petrusz), pp. 155–78. Academic Press, London.

Horisberger, M. and Rosset, J. (1977). Colloidal gold, a useful marker for transmission and scanning electron microscopy. *Journal of Histochemistry and Cytochemistry*, **25**, 295–305.

Horisberger, M., Rosset, J., and Bauer, H. (1975). Colloidal gold granules as markers for cell surface receptors in the scanning electron microscope. *Experentia*, **31**, 1147–9.

Hsu, Y-H. (1984). Immunogold for detection of antigen on nitrocellulose paper. *Analytical Biochemistry*, **142**, 221–5.

Jemmerson, R., Klier, F.G., and Fishman, W.H. (1985). Clustered distribution of human placental alkaline phosphatase on the surface of both placental and cancer cells. *Journal of Histochemistry and Cytochemistry*, **33**, 1227–34.

Kistler, J., Kirkland, B., and Bullivant, S. (1985). Identification of a 70,000-D protein in lens membrane junctional domains. *Journal of Cell Biology*, **101**, 28–35.

Kordeli, E. *et al.* (1986). Evidence for a polarity in the distribution of proteins from the cytoskeleton in *Torpedo marmorata* electrocytes. *Journal of Cell Biology*, **102**, 748–61.

Lackie, P.M., Hennessy, R.J., Hacker, B.W., and Polak, J.M. (1985). Investigation of immunogold–silver staining by electron microscopy. *Histochemistry*, **83**, 545–50.

Lamberts, R. and Goldsmith, P.C. (1985). Pre-embedding colloidal gold immunostaining of hypothalamic neurons. *Journal of Histochemistry and Cytochemistry*, **33**, 499–507.

Lane, B.P. and Europa, D.L. (1965). Differential staining of ultrathin sections of epon-embedded tissues for light microscopy. *Journal of Histochemistry and Cytochemistry*, **13**, 579–82.

Larsson, L-I. (1979). Simultaneous ultrastructural demonstration of multiple peptides in endocrine cells by a novel immunocytochemical method. *Nature*, **282**, 743–4.

Lucocq, J.M. and Roth, J. (1985). Colloidal gold and colloidal silver-metallic markers for light microscope histochemistry. In *Techniques in immunocytochemistry*, Vol. 3 (ed. G.R. Bullock and P. Petrusz), pp. 203–36. Academic Press, London.

Manigley, C. and Roth, J. (1985). Applications of immunocolloids in light microscopy. IV. Use of photochemical silver staining in a simple and efficient double-staining technique. *Journal of Histochemistry and Cytochemistry*, **33**, 1247–51.

Mannweiler, K., Hohenberg, H., Bohn, W., and Rutter, G. (1982). Protein A–gold particles as markers in replica immunocytochemistry: high resolution electron microscope investigations of plasma membrane surfaces. *Journal of Microscopy*, **126**, 145–9.

Mason, D.Y., Cordell, J.L., and Pulford, K.A. (1983). Production of monoclonal antibodies for immunocytochemical use. In *Techniques in immunocytochemistry*, Vol. 1 (ed. G.R. Bullock and P. Petrusz), pp. 175–216. Academic Press, London.

Moeremans, M., Daneels, G., Van Dijck, A., Langanger, G., and De Mey, J. (1984). Sensitive visualisation of antigen–antibody reactions in dot and blot immune overlay assays with immunogold and immunogold–silver staining. *Journal of Immunological Methods*, **74**, 353–60.

Mouton, C. and Lamonde, L. (1984). Immunogold electron microscopy of surface antigens of oral bacteria. *Canadian Journal of Microbiology*, **30**, 1008–13.

Murphy, J.A. (1980). Non-coating techniques to render biological specimens conductive/1980 update. *Scanning Electron Microscopy*, **1**, 209–20.

Orefici, G., Molinari, A., Donell, G., Paradisi, S., Teti, G., and Arancia, G. (1986). Immunolocation of lipoteichoic acid on group B streptococcal surface. *FEMS Microbiology Letters*, **34**, 111 15.

Park, L., Albrecht, R.M., Simmons, S.R., and Cooper, S.L. (1986). A new approach to study adsorbed proteins on biomaterials: immunogold staining. *Journal of Colloids and Interface Science*, **111**, 197–210.

Pinto da Silva, P. (1984). Freeze–fracture cytochemistry. In *Immunolabelling for electron microscopy* (ed. J.M. Polak and I.M. Varndell), pp. 179–88. Elsevier Science Publishers, BV, Amsterdam.

Pinto da Silva, P. and Kan, F.W. (1984). Label–fracture: a method for high resolution labeling of cell surfaces. *Journal of Cell Biology*, **99**, 1156–61.

Pinto da Silva, P., Kachar, B., Torrisi, M.R., Brown, C., and Parkinson, C. (1981).

Freeze–fracture cytochemistry: replicas of critical point-dried cells and tissues after fracture label. *Science*, **213**, 230–3.

Pinto da Silva, P., Barbosa, M.L.F., and Aguas, A.P. (1986). A guide to fracture label: cytochemical labelling of freeze–fractured cells. In *Advanced techniques in biological electron microscopy* (ed. J.K. Koehler), pp. 201–27. Springer-Verlag, Berlin.

Polak, J.M. and Van Noorden, S. (1987). An introduction to immunocytochemistry: current techniques and problems. *Royal Microscopical Society Handbook II* (Revised edn). Oxford University Press.

Robenek, H. and Severs, N.J. (1984). Double labelling of lipoprotein receptors in fibroblast cell surface replicas. *Journal of Ultrastructure Research*, **87**, 149–58.

Robinson, E.N. *et al.* (1984). Ultrastructural localisation of specific gonococcal macromolecules with antibody–gold sphere immunological probes. *Infection and Immunity*, **46**, 361–6.

Romano, E.L. and Romano, M. (1977). Staphyloccocal protein A bound to colloidal gold: a useful reagent to label antigen–antibody sites in electron microscopy. *Immunocytochemistry*, **14**, 711–15.

Romano, E.L., Stolinski, C., and Hughes-Jones, N.C. (1974). An antiglobulin reagent labelled with colloidal gold for use in electron microscopy. *Immunocytochemistry*, **11**, 521–2.

Romano, E.L., Stolinski, C., and Hughes-Jones, N.C. (1975). Distribution and mobility of the A_1 D_1 and C antigens on human red cell membranes; studies with a gold labelled antiglobulin reagent. *British Journal of Haematology*, **30**, 507–16.

Roth, J. (1982a) The protein A–gold (pAg) technique: a qualitative and quantitative approach for antigen localisation on thin sections. In *Techniques in immunocytochemistry*, Vol. 1 (ed. G.R. Bullock and P. Petrusz), pp. 107–33. Academic Press, London.

Roth, J. (1982b). The preparation of protein A–gold complexes with 3 nm and 15 nm gold particles and their use in labelling multiple antigens on ultrathin sections. *Histochemical Journal*, **14**, 791–801.

Roth, J. (1982c). Applications of immunocolloids in light microscopy: preparation of protein A–silver and protein A–gold complexes and their applications for localisation of single and multiple antigens in paraffin sections. *Journal of Histochemistry and Cytochemistry*, **30**, 691–6.

Roth, J. (1983a). The colloidal gold marker system for light and electron microscopic cytochemistry. In *Techniques in immunocytochemistry*, Vol. 2 (ed. G.R. Bullock and P. Petrusz), pp. 216–84. Academic Press, London.

Roth, J. (1983b). Application of lectin–gold complexes for electron microscopic localisation of glycoconjugates on thin sections. *Journal of Histochemistry and Cytochemistry*, **31**, 987–99.

Roth, J. and Binder, M. (1978). Colloidal gold, ferritin and peroxidase as markers for electron microscopic double labelling lectin techniques. *Journal of Histochemistry and Cytochemistry*, **26**, 163–9.

Roth, J., Bendayan, M., and Orci, L. (1978). Ultrastructural localization of intracellular antigens by the use of protein A–gold complex. *Journal of Histochemistry and Cytochemistry*, **26**, 1074–81.

Roth, J., Bendayan, M., and Orci, L. (1980). FITC–protein A–gold complex for light and electron microscopic immunocytochemistry. *Journal of Histochemistry and Cytochemistry*, **28**, 55–7.

Sas, D.F., Sas, M.J., Johnson, K.R., Menko, A.S., and Johnson, R.G. (1985). Junctions between lens fiber cells are labelled with a monoclonal antibody shown to be specific for MP26. *Journal of Cell Biology*, **100**, 216–25.

Schmitt, D., Faure, M., Dezutter-Dembuyant, C., and Tuffery, D. (1984). Ultrastructural immunogold labelling of Langerhans cells enriched epidermal cell suspensions. *Archives of Dermatological Research*, **276**, 27–32.

Schulze, E. and Kirschner, M. (1986). Microtubule dynamics in interphase cells. *Journal of Cell Biology*, **102**, 1020–31.

Slot, J.W. and Geuze, H.J. (1981). Sizing of protein A–colloidal gold probes for immunoelectron microscopy. *Journal of Cell Biology*, **90**, 533–6.

Slot, J.W. and Geuze, H.J. (1984). Gold markers for single and double immunolabelling of ultrathin cryosections. In *Immunolabelling for electron microscopy* (ed. J.M. Polak and I.M. Varndell), pp. 129–42. Elsevier Science Publishers, BV, Amsterdam.

Slot, J.W. and Geuze, H.J. (1985). A new method of preparing gold probes for multiple-labelling cytochemistry. *European Journal of Cell Biology*, **38**, 87–93.

Soligo, D., de Harven, E., Nava, M.T., and Lambertenghi Delilers, G. (1986). Immunocytochemistry with backscattered electrons. In *Science of biological specimen preparation* (ed. M. Mueller, R.P. Becker, A. Boyde, and J.J. Wolosewick), pp. 289–97. SEM Inc., Illinois, USA.

Studer, D. and Herman, R. (1986). Colloidal gold particles detected on highly structured surfaces of large samples by backscattered electrons in the scanning electron microscope. In *Science of biological specimen preparation* (ed. M. Mueller, R.P. Becker, A. Boyde, and J.J. Wolosewick), pp. 203–7. SEM Inc., Illinois, USA.

Taatjes, D., Ackerstrom, B., Bjorck, L., Carlemalm, E., and Roth, J. (1987). Streptococcal protein G–gold complex: comparison with staphylococcal protein A–gold complex for spot blotting and immunolabelling. *European Journal of Cell Biology*, **45**, 151–9.

Tapia, F.J., Varndell, I.M. Probert, L., De Mey, J., and Polak, J.M. (1983). Double immunogold staining method for the simultaneous ultrastructural localisation of regulatory peptides. *Journal of Histochemistry and Cytochemistry*, **31**, 977–81.

Tetley, L., Turner, C.M.R., Barry, J.D., Crowe, J.S., and Vickerman, K. (1987). Onset of expression of the variant surface glycoproteins of *Trypanosoma brucei* in the tsetse fly studied using immunoelectron microscopy. *Journal of Cell Science*, **87**, 363–72.

Tokuyasu, K.T. (1978). A study of positive staining of ultrathin frozen sections. *Journal of Ultrastructure Research*, **63**, 287–307.

Tokuyasu, K.T. (1984). Immuno-cryoultramicrotomy. In *Immunolabelling for electron microscopy* (ed. J.M. Polak and I.M. Varndell), pp. 71–82. Elsevier Science Publishers, BV, Amsterdam.

Tokuyasu, K.T. (1986). Immunocryoultramicrotomy. *Journal of Microscopy*, **143**, 139–49.

Tolson, N.D., Boothroyd, B., and Hopkins, C.R. (1981). Cell surface labelling with gold colloidal particles: the use of avidin and staphylococcal protein A-coated gold in conjunction with biotin and Fc-bearing ligands. *Journal of Microscopy*, **123**, 215–26.

Van Noorden, S. and Polak, J.M. (1985). Immunocytochemistry of regulatory peptides. In *Techniques in immunocytochemistry*, Vol. 3 (ed. G.R. Bullock and P. Petrusz), pp. 115–154. Academic Press, London.

Varndell, I.M. and Polak, J.M. (1984). Double immunostaining procedures: techniques and applications. In *Immunolabelling for electron microscopy* (ed. J.M. Polak and I.M. Varndell), pp. 155–77. Elsevier Science Publishers, BV, Amsterdam.

Walker, P.D. and Beesley, J.E. (1985). Trends in the localisation of bacterial antigens by immunoelectron microscopy. *Annals of the New York Academy of Sciences*, **420**, 410–21.

Walther, P. and Muller, M. (1986). Detection of small (5–15 nm) gold-labelled surface antigens using backscattered antigens. In *Science of biological specimen preparation* (ed. M. Mueller, R.P. Becker, A. Boyde, and J.J. Wolosewick), pp. 195–201. SEM Inc., Illinois, USA.

Wang, T.L. (1986). *Immunology in plant science.* S.E.B. Seminar Series, No. 29. Cambridge University Press, Cambridge.

Wehland, J. and Willingham, M.C. (1983). A rat monoclonal antibody reacting specifically with the tyrosylated form of alpha-tubulin. II. Effects on cell movements, organisation of microtubules and intermediate filaments and arrangement of Golgi elements. *Journal of Cell Biology*, **97**, 1476–90.

Weibel, E.R. (1969). Stereological principles for morphometry in electron microscopic cytology. *International Review of Cytology*, **26**, 235–302.

Wolosewick, J.J., de Mey, J., and Meininger, V. (1983). Ultrastructural localisation of tubulin and actin in polyethylene glycol-embedded rat seminiferous epithelium by immunogold staining. *Biologie Cellulaire*, **49**, 219–26.

Yang, H.Y., Leiska, N., Goldman, A.E., and Goldman, R.D. (1985). A 300,000-mol-wt intermediate filament-associated protein in baby hamster kidney (BHK-21) cells. *Journal of Cell Biology*, **100**, 620–31.

Yoshimura, N., Murache, T., Heath, R., Kay, J.J., Jasani, B., and Newman, G.R. (1986). Immunogold electron microscopic localisation of calpain I in skeletal muscle of rats. *Cell and Tissue Research*, **244**, 265–70.

Index